T. Blackburn

Notes on Australian Coleoptera with descriptions of new species

T. Blackburn

Notes on Australian Coleoptera with descriptions of new species

ISBN/EAN: 9783741183911

Manufactured in Europe, USA, Canada, Australia, Japa

Cover: Foto ©berggeist007 / pixelio.de

Manufactured and distributed by brebook publishing software
(www.brebook.com)

T. Blackburn

Notes on Australian Coleoptera with descriptions of new species

NOTES ON AUSTRALIAN *COLEOPTERA* WITH DESCRIPTIONS OF NEW SPECIES.

By the Rev. T. Blackburn, B.A., Corr. Mem. Linn. Soc. N.S.W.

The following memoir, like others that I have written under a similar title, is of a somewhat miscellaneous character. The very limited amount of leisure time at my disposal and the shortness of the intervals in which it occurs, render it very difficult for me to make an exhaustive study of an individual family, or even genus; and I have usually to be content with merely noting from time to time the characters of new genera or species that may come haphazard into my hand, observations on synonymy, and aught else that may appear to call for publication. This I trust will be accepted as an apology for the too miscellaneous contents of the memoir.

I have the greater pleasure in offering the memoir to the Linnean Society, because it gives me the opportunity of expressing in the pages of their Proceedings my acknowledgments of the assistance I have received in its preparation through the courtesy of the Hon. William Macleay, who has been good enough to compare with his types some South Australian forms, and so enable me to feel confident that I am working in harmony with the many valuable publications on Australian *Coleoptera* that he has put forth.

CARABIDÆ.

SILPHOMORPHA SPRETA, sp.nov.

Lata; depressa; minus nitida; supra nigra; labro, mandibulis, antennis, et marginibus lateralibus rufescentibus; elytris magnâ parte flavis ; subtus (capite nigro excepto), cum pedibus, rufo-picea ; elytrorum epipleuris basi intus late flavis.

[Long. 5 lines, lat. 2½ lines.

The lateral margins of the prothorax and elytra are narrowly, obscurely reddish pitchy ; each elytron bears a large yellow spot which almost reaches both base and apex, and is separated from the lateral margin by only about a fifth of the width of the elytron, its outline next the suture being deeply emarginate ; if the yellow color be regarded as the ground tint of the elytra they would appear to be margined rather widely at the side, and very narrowly in front and behind, with black, and to bear a large common black mark which (viewed with the head of the insect towards the observer) has the appearance of an open umbrella standing up on its handle, the widest part of this mark extending about half-way across each elytron. The head is wide, and short in front of the eyes ; the clypeus is rather deeply emarginate ; on either side of the head a well-defined oblique furrow runs (a little within the eye) from the front nearly to the base,— these furrows converging but not nearly meeting hindward. The surface of the head and prothorax is evenly and finely coriaceous. The latter is slightly and undefinedly uneven on the disc with its front margin strongly bisinuate and its base nearly straight. The elytra have rather wide, and rather decidedly turned up, lateral margins ; their surface is coriaceous uniformly with the head and prothorax, and bears also about eight rows of very faintly impressed punctures. The underside is pitchy much variegated with reddish, the darkest parts being the head, the tibiæ and tarsi, the epipleuræ of the prothorax, the same of the elytra down their middle part, and the hinder part of the hind body. The conspicuous bright yellow color of the inner part of the base of the epipleuræ of the elytra is a striking character.

This species must resemble *S. marginata*, Castln., from the Paroo River, the description of which however (beyond the words " broad, depressed ") deals only with color. In that species the elytra are said to be yellow at the base, and the comparison of the markings to those of *nitiduloides*, Guér., implies (rather vaguely) that the yellow color almost touches the lateral margins. The decription, however, is so vague that *S. spreta* may possibly not resemble it very much.

Northern Territory of S. Australia; collected by Mr. J. P. Tepper ; a female specimen.

S. BOOPS, sp.nov.

Supra colore variabilis (ferruginea plus minus infuscata, vel tota picea) ; corpore subtus antennis palpis pedibusque ferrugineis ; capite prothoraceque vix evidenter punctulatis, illo lateribus sub-depressis, oculis prominentibus ; elytris plus minus punctulato-striatis. Long. 6 lines, lat. 3½ lines.

This species belongs to the very difficult and obscure group in the genus of which *S. fallax*, Westw., is a member. Mr. Macleay tells me that it is distinct from his *S. Mastersi*, which is said to be more strongly punctulate-striate on the elytra than *S. fallax*. The present insect is evidently a very variable one ; the series before me varies in the color of the upper surface from nearly uniform ferruginous to nearly uniform piceous, and in the sculpture of the elytra from being nearly smooth to being as strongly punctulate-striate as a typical *S. fallax*. The head and prothorax scarcely differ from the same parts in *S. fallax*, except in the less con-vexity of the former (especially at the sides), which makes the eyes appear more prominent; and in the latter being wider in proportion to the length of the whole insect, its width being scarcely less than half the whole length. The elytra are decidedly wider in proportion to their length than those of *S. fallax*, being scarcely more than an eighth again (in *fallax* they are about a quarter again) as long as together wide, and the lateral margins are wider, and decidedly more (though still not at all strongly) rounded than in that species.

Northern Territory of S. Australia; collected by Mr. J. P. Tepper.

GNATHAPHANUS DARWINI, sp.nov.

Niger, vix ænescens ; antennis, palpis, pedibusque plus minusve rufescentibus ; prothorace antice minus angustato, angulis posticis rotundato-rectis ; elytrorum interstitiis 3° (duplici serie) 5° et 9° seriatim punctulatis. Long. 3½ lines, lat. 1½ lines.

The front of the labrum, the palpi, and the basal joint of the antennæ are brownish testaceous, the remaining joints of the latter being of the same color more or less marked with piceous; the legs are brown. The surface of the head is even behind the clypeal suture except that there is a puncture in a feeble depression on either side in front, and another near the inner margin of each eye. The prothorax is not quite half again as wide as it is long down the middle, its front margin not much narrower than its base ; its sides are moderately rounded, the flattened margin being very narrow in front, but widening considerably hindward ; the dorsal channel is well marked, the arched impression in front fairly defined, the basal impression on either side shallow, but not small; the surface is devoid of punctures, except the setiferous one on either lateral margin. The elytra are rather strongly striated, the abbreviated stria rather long and well-defined ; the interstices are flat ; the 3rd interstice bears four large punctures in its front half and two (far apart) in its hinder half close to its outer edge, and also (in its hinder half) close to its inner edge several similar ones ; the 5th interstice bears six or seven (similar) close to its outer edge, and the 9th a somewhat more numerous series interspersed with some smaller punctures.

The double series of punctures on the 3rd interstice seems to distinguish this species from all its allies.

Northern Territory of South Australia ; collected by Mr. J. P. Tepper.

HYPHARPAX PARVUS, Chaud.

I have very little doubt that this is identical with *Harpalus* (*Hypharpax*) *inornatus*, Germ.

CRATOGASTER MELAS, Cast.

The Baron de Chaudoir (Ann. Mus. Gen. Vol. VI., p. 574) has stated that he has examined the original type of this insect, and found it to be identical with *Feronia (Cratogaster) sulcata*, Blanch. It, therefore, appears to be by an oversight that the two stand in Mr. Masters' Catalogue as distinct.

RHYTISTERNUS SULCATIPES, sp.nov.

Sat depressus; niger; antennis palpis tarsisque elongatis gracilibus plus minus rufescentibus; prothorace vix transverso, postice utrinque bistriato,—striis haud in excavatione perspicua positis, lateribus postice haud sinuatis; elytris striis 5, 6, et 7-a plus minus obsoletis; tarsis posticis extus sat fortiter sulcatis. Long. 7 lines, lat. $2\frac{2}{5}$ lines.

A depressed parallel somewhat slender insect, having much the build of the European *Adelosia picimanus*, Heer. The antennæ and tarsi are longer and much more slender than those of *R. liopleura*, Chaud.; the palpi also are more slender, otherwise the head does not differ noticeably. The prothorax is scarcely a quarter as wide again as it is long down the middle, its base scarcely narrower than its front margin, which is slightly concave, with anterior angles scarcely produced; the sides are moderately and rather evenly rounded, but so that the prothorax is at its widest just in front of the middle (though less in front of the middle than that of *R. liopleura*); they are not at all sinuate behind; the hind angles are very obtuse, but not quite rounded, and not in the least "subdentate" (as they are in *R. liopleura*, Chaud.); the transverse impression on the front of the disc is scarcely marked, the dorsal channel fairly strong but not very nearly reaching either the base or the front margin; there are two well-defined sulci on either side behind, which are not placed in an excavation, but are separated by a space continuous with the general surface of the prothorax; of these the inner one is linear and sharply cut, the outer more obscure and foveiform. On the elytra the 5th, 6th, and 7th striæ from the suture are progressively fainter than the preceding four, so that the 7th (though traceable throughout with a good lens) is *extremely* faint; in the apical 6th part of the elytra these are as strongly marked as the inner striæ, the 5th being at the extreme base also not much feebler than the 4th. The clearly-defined sulcus on the external side of the hind tarsi is a conspicuous character. The prosternum is not margined between the anterior coxæ. None of the elytral interstices except the 9th are convex unless slightly so close to the apex. The puncturation of the 9th interstice resembles the same in *R. liopleura*. The sculpture of the prosternal episterna is very strong.

The species of this genus (or subgenus) are very close to each other and difficult to identify. The Baron de Chaudoir has described one as having the sides of the prothorax not sinuate behind (*R. liopleura*), and named three other species without describing them (merely pointing out some differences from *liopleura* in respect of two, and from one of those two in respect of the third). The Count de Castlenau has also described most of these under different names from those of de Chaudoir, but his descriptions are of little value, the type of the genus which he used for comparison with other species having been stated by de Chaudoir to have been wrongly named. The insect I have just described differs from the description of *liopleura* (and from specimens which I believe to be that insect) in having the hind angles of the prothorax not in the smallest degree dentate, and the two basal sulci of the same not placed in an excavation, in the external sulcation of the hind tarsi (de Chaudoir gives as a generic character "hind tarsi *generally* not sulcate," and does not mention *liopleura* as forming an exception), and in the very much more slender antennæ and tarsi. From *R. lævilatera, R. sulcatipes* differs *inter alia* in not having the 5th, 6th, and 7th striæ on the elytra "altogether obliterated;" from *R. cyathodera* in not having the thorax "much wider and shorter (than that of *liopleura*);" and from *R. misera* in being very much larger, with elytra differently striated, &c.

I may say that Baron de Chaudoir's description of *R. liopleura* appears to me faulty in calling the sides of the prothorax "not" sinuate behind, and the 7th elytral stria "altogether obliterated." The former would be better described as "scarcely sinuate," and the latter as "almost obliterated.' I have no doubt of the correctness of this emendation, because the baron states that specimens ticketed *F. Australasiæ* in Castelnau's collection (which species Castelnau speaks of as having the sides of the prothorax sinuate behind) are identical with his *liopleura;* moreover, Castelnau states that this insect (his *F. Australasiæ*, Dej., but, according to Chaudoir, not really that species) is common in South Australia; and there is a *Rhytisternus*, the only one common in South Australia, well known to me, which I had purposed describing as new until I noticed this discrepancy between de Chaudoir's and de Castelnau's descriptions, but which I now have no doubt is the species that de Chaudoir described as *liopleura*, and that de Castelnau called *Feronia Australasiæ*, Dej., and it has the sides of the prothorax slightly sinuate behind, and the 7th elytral stria, though excessively faint, yet certainly traceable with a good lens.

I have specimens of the insect described above from the neighbourhood of Adelaide and from Yorke's Peninsula; it appears to be rare.

Pristonychus Australis, sp.nov.

Minus convexus; subnitidus; niger vel piceo-niger; antennis palpis tarsisque rufo-piceis; elytris subcyaneis; prothorace postice vix angustato, angulis posticis obtusis, subdentiformibus; elytris striatis, striis subtiliter punctulatis; tarsis sat brevibus.
[Long. 7 lines, lat. $2\frac{2}{5}$ lines.

The head and prothorax are nitid, the former with a strong longitudinal sulcus on either side between the eyes. The prothorax is about a quarter again as wide as long, the front margin and base nearly equal in width, the former slightly concave, the latter gently bisinuate; the sides are gently arched in front, and lightly sinuate behind the middle; the front angles are feebly defined and rounded, the hind angles obtuse but nearly right angles, with the extreme apex however rounded off, and directed somewhat outward; on the surface the anterior transverse impression is strong, the dorsal channel well defined but reaching neither the front margin nor the base, and the basal fovea on either side elongate and extremely deep. The elytra are at their widest behind the middle, their sides gently rounded; they are moderately strongly striated, the striæ finely and neither closely nor very noticeably punctulate, the interstices somewhat convex in front, but scarcely so behind; the abbreviated stria near the scutellum on each elytron is long (about equal to the basal two joints together of the antennæ) and very deep, almost foveiform at the apex,—the suture between them being more elevated than behind. The eighth stria bears an irregular row of rather large punctures.

Resembles the European *P. subcyaneus*, Illig., but differs from it *inter alia* as follows :—the prothorax is much less narrowed behind with the hinder part of its sides much less strongly sinuate, the striation of the elytra is feebler, the tarsi are less elongate and not quite so hairy on the upper surface, and the claws are slightly more crenulate on their inner edge.

Port Lincoln, Wallaroo, and near Roseworthy.

CYBISTER GRANULATUS, sp.nov.

♀Ovalis; depressus; posterius conspicue latior; minus nitidus; supra olivaceus, capite antice prothoracisque lateribus testaceis; elytris crebrius granulatis, margine externo (cum epipleuris) sat late testaceo; subtus piceus, metathoracis episternis abdominisque lateribus testaceo-maculatis; pedibus 4 anterioribus testaceis, femoribus anterioribus fusco-maculatis, tibiis intermediis et tarsis anterioribus fusco-testaceis, tarsis intermediis pedibusque posticis piceis; antennis testaceis.　　　Long. 12 lines, lat. 7 lines (vix).

This species is colored almost exactly as the common widely distributed *C. tripunctatus*, Ol., save that the front tarsi are darker, and that the dark color of the hind part of the head is rather strongly and angularly produced in the middle into the testaceous color of the front. Its shape, however, is very different, being very much less convex than that of *C. tripunctatus*, with the elytra more dilated behind,—at the widest part behind the middle these are considerably more than half again as wide as at the base,—their sides running in an almost straight line, absolutely from the base, to the widest part and the lateral vitta being of quite even width from base to apex. The head, prothorax, and basal portion of elytra bear a few long irregular scratches. The rows of punctures on the elytra scarcely differ from the same in *C. tripunctatus*, but the general surface of the elytra is more closely and evidently coriaceous so as to appear much less nitid, and is evenly and rather closely studded with small elevated round pustules which, however, become gradually smaller and less elevated from the base to the apex. Immediately within the inner line of punctures, these pustules run in two even parallel rows from the base to the apex, having a narrow smooth space between them. There is some indication of a similar sculpture immediately within the external row of punctures, but these outer rows of pustules leave their line of punctures near the front and bend obliquely across to the base of the inner rows.

Northern Territory of South Australia; taken by Professor Tate.

PALPICORNES.

STERNOLOPHUS TENEBRICOSUS, sp.nov.

Convexus; minus elongatus; nitidus; niger; antennis, labro antice, palpisque labialibus, rufo-testaceis; supra subtilissime crebre punctulatus; capite prothorace et elytris punctis majoribus seriatim positis (his capillos subtiles ferentibus) instructis; prothorace antice æqualiter emarginato, subtus subtiliter crebre punctulatus, crebre breviter pubescens.

[Long. 5½ lines, lat. 2½ lines (vix).

The basal joint of the maxillary palpi is testaceous, the 2nd and 3rd nearly black, with the extreme apices paler, the 4th reddish pitchy ; the anterior tarsi are reddish pitchy. Apart from generic characters, size and color, this insect is so extremely similar to *Hydrobiomorpha Tepperi* that the description of that species will suffice for it, subject to the following remarks :—the prothorax is evenly emarginate in front, without any bisinuation ; the punctures forming the series on the head, prothorax, and elytra are evidently finer ; and the elytral series are not so regular, the inner four appearing to consist each of two closely adjacent rows confused together, while the series close to the lateral margin is wanting. The general form also is less elongate.

Probably this insect is allied to *S. nitidulus*, Macl., the description of which is rather brief ; but that species is said to have the palpi red and to have "faint traces of a few rows of punctures on the elytra." In *S. tenebricosus* the rows are perfectly well-defined and conspicuous.*

A single example taken near Palmerston, N.T., by Mr. J. P. Tepper.

HYDROBIOMORPHA, gen.nov. (HYDROPHILIDÆ).

Mentum antice leviter rotundatum haud sinuatum, angulis anticis vix emarginatis.

Mandibula apice bilobata.

---

* Since writing the above I have found in the South Australian Museum a specimen (in wretched condition) which is probably *S. nitidulus*, Macl. It is extremely like *S. tenebricosus*, but differs in having the maxillary palpi red, and the sculpture of the elytra faint and running more regularly in single rows. The sternal spine also differs ; in both species it reaches nearly to the apex of the first ventral segment and is pointed behind, but in *tenebricosus* the point forms the apex of the *lower* edge of the spine (*i.e.*, that nearest to the surface of the body), so that the upper outline of the carina viewed from the side is declivous at the extreme apex ; while in the other species this is reversed and the point forms the apex of the upper edge of the spine, so that the upper outline of the carina viewed from the side is straight to the apex.

Prosternum (ut in gen. *Hydroo*) carinâ elevatâ postice spinosâ instructum.

Tarsi postici (nec intermedii) vix remiformes. Maris palpi maxillares fortiter dilatati, feminæ gracillimi.

The insects for which I propose this name have very much the appearance of *Hydrobius* at the first glance. They seem to be to some extent intermediate between M. Lacordaire's subfamilies *Hydrophilides* and *Hydrobiides*, having the continuous sternal keel (free at the apex, which is about level with the hindmost edge of the hind coxæ) of the former, with hind tarsi approaching the latter in structure (being narrower and less distinctly remiform than the intermediate tarsi). The following are the leading characters of the genus :—maxillary palpi of male with the joints (especially the third) dilated, of female very long and slender, their second and 3rd joints nearly equal, the 4th only a little shorter; prothorax in front very strongly bisinuate; antennæ very peculiar, 9-jointed, the basal 5 not much different from the same in *Hydrophilus*; the 6th joint smooth and shining like the preceding, but forming a kind of saucer on which the 7th joint is laid in such fashion that very little of the 6th joint can be seen from above, and very little of the 7th from beneath ; the 7th joint is almost exactly of the shape of the bone of a chicken known as the "merry-thought ;" it (as well as the following two) is opaque and pubescent, and ciliated with long golden hairs ; the 8th joint is attached to the apex of the thicker lobe (which lies flat on the saucer-like 6th joint) of the 7th ; it (the 8th) is very short, and very strongly produced in an upward direction so as almost to meet the apex of the thinner lobe of the 7th ; the apical joint is an arched transverse plate, its upper surface the concave one. The hind body is roundly (not, as in *Hydrophilus*, angularly) convex down the middle line. The mandibles end in two lobes, the external one much the shorter and longer ; as far as I can see (without dissection) they are not toothed within.

The carinated prosternum, and claws dentate at their base on all the legs, distinguish this genus from *Tropisternus ;* the latter character distinguishes it from *Hydrous* and *Sternolophus*.

H. BOVILLI, sp.nov.

Minus convexa; sat elongata; nitida; nigra; clypeo labroque antice rufis; palpis, antennis (in media) et tarsis rufescentibus; supra subtiliter sat crebre punctulata; capite prothorace et elytris punctis majoribus seriatim positis (his capillos subtiles ferentibus) instructis; subtus subtiliter crebre punctulata, crebre breviter pubescens.   Long. 7½ lines, lat. 3 lines.

The head is moderately wide (across the eyes about two-thirds the width of the prothorax); the red anterior margin of both clypeus and labrum is very conspicuous; the puncturation (which covers it and the rest of the upper surface very evenly) is about equally fine with that of the same parts in the European *Hydrous caraboides*, L., but is quite evidently less close; the larger puncturation is as follows :—a pair of punctures placed transversely in front, and about five placed in a transverse row on either side of the base of the dark part of the clypeus; and on either side an elongate cluster just within the eye, and another curving in a half circle forward from just in front of the eye. The prothorax is at the base quite twice as wide as it is long down the middle, and about half again as wide as its front margin; its anterior angles are well advanced and rather sharp, its hind angles roundly rectangular, and its lateral margins nearly straight; the large punctures run in two series on either side obliquely backward from near the front and about the middle of the lateral margin. The elytra are about three and a half times longer than the prothorax, truncate at the base, very gently and arcuately contracted hindward, with humeral angles little marked; the fine lateral margin is continued along the base on either side to the scutellum; the rows of larger punctures run as follows—two rows (very little larger than those of the general surface) near the scutellum on either side, the inner of which in front forks forward into two branches at about twice the distance of the scutellum from the base, and does not reach the apex though both its branches reach the base; then three rows of punctures about equal in size to those in the prothoracic series, followed by two rows rather close together near the margin ; of these rows the 1st and 3rd are not continuous near the base, and all (especially the lateral ones) are somewhat irregular through some of the punctures being out of line.

A single specimen of this very interesting insect has been sent to me by Dr. Bovill, who took it near Palmerston, in the Northern Territory of South Australia.

H. TEPPERI, sp.nov.

Sat convexa; sat elongata; nitida; nigra, clypeo labroque antice rufis; palpis, antennis (articulis ultimis exceptis), tarsis, non nullis exemplis femoribus etiam, rufescentibus ; supra subtiliter sat crebre punctulata; capite prothorace et elytris punctis majoribus seriatim positis (his capillos subtiles ferentibus) instructis ; subtus subtiliter crebre punctulata, crebre breviter pubescens.

[Long. 7 lines, lat. 3 lines.

Decidedly more convex than the preceding, and much more parallel-sided, the elytra being quite as wide a little behind the middle as at the base ; the prothorax is not so strongly bisinuate in front ; the puncturation of that segment and of the head scarcely differs from the same in *Bovilli* ; the sculpture of the elytra is very different, as the rows of punctures intermediate in size between the uniform surface punctures and the five rows of much larger punctures are altogether wanting. In all other respects the description of the former insect would apply to this one.

Palmerston, N.T. (Mr. J. P. Tepper) ; also Yam Creek, N.T. (Prof. Tate).

HYDROBIUS.

According to Dr. Sharp (Trans. Ent. Soc., 1884) *H. assimilis*, Hope, which was founded on a specimen from Port Essington, is a common Australian species and is identical with *H. Zealandicus*, Broun (from New Zealand). I am well acquainted with a common and widely distributed (in Southern Australia) species that is apparently not separable from the New Zealand insect. Dr. Sharp does not *say* that he has examined Hope's type, but I have little doubt he has done so, as the description would hardly suggest the idea of identity with *H. Zealandicus*. I have seen several · collections from the Northern Territory containing *Palpicornes*, in none of which were there any examples of the common Southern Australian species, although I have seen a species from that locality very different from it which appeared to me not unlikely to be *H. assimilis*. Assuming Dr. Sharp to be right in his identification (as I have no doubt he is) I offer the following description of *H. assimilis*, Hope, as likely to be interesting to Australian students, who certainly are not likely to identify the insect on Hope's description.

H. ASSIMILIS, Hope.

Nitidus; minus elongatus; piceo-niger; antennis palpis pedibusque rufescentibus; abdomine rufo-maculato; crebre subtilius punctulatus et punctis majoribus seriatim instructus; elytrorum interstitiis planis; subtus crebre breviter pubescens.

[Long. 4½ to 5 lines, lat. 2¼ to 2¾ lines.

The fine evenly-distributed puncturation of the upper surface is decidedly finer than in the European *H. fuscipes*, Linn.; a row of larger punctures runs across the labrum, another (arched and interrupted in the middle) across the clypeus, and a third curves round the inner margin and front of each eye. On either side of the prothorax two similar lines run from the margin inwards,—one in front of, the other behind, the middle. On the elytra a sutural stria commences faintly about the middle and runs back deepening to the apex: outside this there are nine rows of punctures similar to those of the thoracic series which are obsolete in front (especially those near the suture), but become strongly defined behind; between the 1st and 2nd, 3rd and 4th, 5th and 6th, 7th and 8th of these, and outside the 9th, there is in each case an irregular row of still larger punctures. The red marks on the hind body are not at all conspicuous and consist of

a small spot on either side close to the margin on each of the basal
four segments. The metasternum is roundly convex as in
H. fuscipes; the meso- and pro-sterna are acutely carinate as in
H. oblongus, Hbst., moderately long cilia springing from each
carina. The general form is longer and more parallel than in
H. fuscipes, with the anterior angles of the prothorax less defined.

A variable insect. The following is, so far as I have seen, about
the extreme of its variety :—a little smaller thar the type ; color
a deeper black ; no red markings on the underside ; the puncturation throughout a little less close ; the rows of punctures better
defined, all being clearly traceable to the base ; the interstices
quite strongly convex in their posterior third part. It is possible
this may be a good species.

Common in South Australia.

H. MACER, sp.nov.

Nitidus ; angustus; convexus; elongatus; olivaceo-niger; antennis
palpis pedibusque rufis ; prothorace elytrisque anguste testaceo-
marginatis ; crebre subtiliter punctulatus et punctis majoribus
seriatim instructus ; elytrorum interstitiis planis ; subtus crebre
breviter pubescens, piceo-ferrugineus, obscure rufo-maculatus.

[Long. 4 lines, lat. 1⅓ lines.

The description of *H. Australis* might be read to apply to this
species in all respects except color and shape, the sculpture of the
segments presenting no noticeable difference. It is, however, a
notably narrower, more convex, and more parallel insect, with
elytra very little less than twice as long as together they are wide,
while those of *Australis* are scarcely half again as long as wide.

A single specimen in my collection, from Victoria; exact locality
not known.

PARACYMUS.

Dr. Sharp (loc. cit.) mentions that he has seen in the collection
made by the Count de Castelnau examples of *Hydrobius (Paracymus) nitidiusculus*, Broun (a species described on New Zealand

53

specimens), which came from Australia. I have met with a species which agrees very fairly with Captain Broun's description, and as the insect has not yet been described in any Australian publication, the following will probably be of interest to Australian readers. I may add that I can discover only eight joints in the antennæ of this and the two following species, which would associate them with *P. æneus*, Germ., a species for which Dr. Sharp has pointed out in the Ent. M. Mag. (Vol. xxi., p. 112), that a new generic name may be necessary on account of this character. It should be noted also that the two species I have named *Lindi* and *sublineatus* will probably eventually be considered generically distinct from all their allies yet described, since they differ from *Paracymus* in the tendency of the elytral puncturation to run in rows, in the absence of a prosternal keel, and in the shape of the mesosternal keel, which is very peculiar indeed; on the front half of the mesosternum it is non-existent, but in the hinder (almost perpendicular) portion the external margin seems to be formed by a keel which also runs round the rather wide base, and emits from the middle of the latter a central keel which runs forward (down the declivity of the mesosternum) for a short distance. Hence, viewed from above the mesosternum seems to rise from the general surface (not as in *Paracymus* as a sharp point, but) in the form of a transverse ridge, which (on account of the convexity of the mesosternum) is of a curved shape, its convex side being turned forward. The general resemblance to *Anacæna* as well as the inclination towards that genus of the structural peculiarities suggests the probability that other forms intermediate between *Paracymus* and *Anacæna* may yet be discovered; I therefore think it better for the present to regard these insects as forming merely a section of *Paracymus*; to which for convenience of reference the name *Paranacæna* might suitably be applied.

P. NITIDIUSCULUS, Broun.

Breviter oblongus; sat convexus; nitidus; supra æneus; antennis (clavâ piceâ exceptâ), palpis, marginibus lateralibus, et pedibus, plus minus rufescentibus; æqualiter minus fortiter, minus crebre, punctulatus; elytris (striâ suturali antice abbreviatâ exceptâ) haud striatis; subtus niger, subtiliter coriaceus, brevissime pubescens.   Long. 1⅓-1⅔ lines, lat. ⅔ line.

There is a considerable variation in the distinctness of the ferruginous tone of the lateral margins of the prothorax and elytra, and in the color of the legs, which are almost black in some specimens.   This insect scarcely differs specifically from the European *Paracymus nigro-æneus*, Sahl., except in respect of its puncturation, which is considerably finer and scarcely so close.

Appears to be common in South Australia; I have it, or have seen it, from Port Lincoln, York's Peninsula and various localities near Adelaide; I have taken it in Western Victoria also.

P. (PARANACÆNA) LINDI, sp.nov.

Breviter oblongus; sat convexus; nitidus; supra nigro-fuscus; capite ad latera ante oculos, palpis, antennis, marginibus lateralibus, et pedibus plus minus dilutioribus; capite subtilius crebre, prothorace sparsius etiam subtilius, elytris fortius vix sublineatim nec crebre, punctulatis; his (stria suturali antice abbreviata excepta) haud striatis; subtus niger, subtiliter coriaceus, brevissime pubescens.   Long. 1⅔ lines, lat. ⅔ line.

The color varies from that above described to an almost uniform pale brown with the paler parts nearly testaceous.   Extremely like the European *Anacæna variabilis*, Shp., in general appearance but differing from it, *inter alia*, in the evidently stronger and more sparing puncturation of the elytra and in the tendency (evident though slight) of the same to run in rows.

Port Lincoln.

P. (PARANACÆNA) SUBLINEATUS, sp.nov.

Breviter oblongus; sat convexus; nitidus; supra niger; antennis (clava excepta), palpis, marginibus lateralibus, et pedibus, plus minus rufescentibus; capite prothoraceque vix evidenter, elytris sublineatim minus fortiter nec crebre, punctulatis; his (stria suturali antice abbreviata excepta) haud striatis ; subtus niger, subtiliter coriaceus, brevissime pubescens.

Long. 1⅓ lines, lat. ⅔ line (vix).

Not very much like any other species known to me. Its general appearance at the first glance is much that of an *Anacæna*, but on closer inspection the very feebly punctured head and prothorax, and the very evident tendency of the elytral puncturation to run in rows, give it a distinctive character among its allies. I have seen only a single specimen, and have little doubt that a long series would show as much color variation as in the preceding.

Roseworthy, S. Australia.

PHILHYDRUS LÆVIGATUS, sp.nov.

Ovalis; nitidus; brunneus, capite obscuriore, prothoracis disco et elytrorum sutura infuscatis ; antennis palpisque testaceis ; his apice vix infuscatis ; capite prothoraceque subtilissime, elytris subtiliter, punctulatis ; subtus niger, pedibus rufis, femoribus vix infuscatis   Long. 1½ lines, lat. ¾ line.

The head is of a dark pitchy color, the clypeus (especially at the sides) paler ; the prothorax has a large obscure fuscous cloud in the middle of the disc ; the elytra are infuscate along the suture and at a short distance within the margins. In size, shape and coloring of elytra, this species resembles the European *P. marginellus*, Fab., but the head and prothorax (as also the palpi, antennæ, and legs) are quite differently colored, and the puncturation of all parts is very much finer ; the puncturation (especially that of the head and prothorax) can scarcely be discerned at all under a less powerful lens than a Coddington. There is no indication whatever of any striæ on the elytra except the sutural one, which is wanting in the anterior third part.

I took a single specimen in Western Victoria ; there is also a specimen in the South Australian Museum, taken by Mr. Tepper at Border Town, in which the ground color is paler but more suffused with brownish, so that the dark suture is less conspicuous.

HYDROBATICUS AUSTRALIS, sp.nov.

Ovalis; minus convexus; minus nitidus; brunneus, fuscoumbratus; prothorace antice quam postice sat evidenter angustiori ; crebre sat fortiter duplo-punctulatus ; elytris obscure striatopunctulatis.   Long. 2½ lines, lat. 1⅓ lines.

Head almost wholly testaceous ; clypeal suture well defined ; front of clypeus decidedly emarginate ; surface of head strongly and not very closely punctured ; labrum black ; palpi and antennæ testaceous, the latter with the club dusky. Prothorax considerably wider than long (as 5 to 3), narrowed from base to apex with gently curved sides ; base about half again as wide as apex, the former nearly straight, the latter slightly emarginate ; hind angles slightly marked, obtuse ; front angles quite rounded off ; surface reddish testaceous with some fuscous markings the most conspicuous of which are two longitudinal lines placed in the hinder half one on either side of the middle, closely and rather strongly punctured (the punctures of different sizes confusedly mixed together). Elytra rounded at the apex, each with about 10 rows of punctures placed in scarcely impressed striæ, the punctures very closely packed in the rows ; the interstices quite flat and confusedly studded with punctures similar to those in the rows, and also (in about equal numbers) with much smaller punctures. The underside is black, very minutely and closely punctured ; the femora are black, the tibiæ and tarsi reddish.

Somewhat variable in color. I have seen specimens, which I cannot separate specifically, with the legs and the whole upper surface testaceous except the two dark lines on the prothorax which are equally well defined in all the specimens I have seen, and therefore more conspicuous in the lighter-colored examples.

Apparently common throughout South Australia ; I have it also from Victoria.

N.B.—This insect no doubt resembles *H. tristis*, Macl., and *luridus*, Macl., (from Queensland), but differs in having the thorax decidedly wider at the base than in front, and doubtless in other particulars. Mr. Macleay has done me the favor of looking at this insect and informing me that it is distinct from the two he has described.

HYGROTOPHUS NUTANS, Macl.

I should say that M. Fairemaire is quite mistaken in thinking that his *Berosus externespinosus* is identical with this insect. I have specimens before me which I believe to be *H. nutans*. From the description it would appear to be a much smaller insect than *B. externespinosus*, and covered with pubescence. If I am right in my identification I cannot regard the verticality of the head or the rounded basal outline of the thorax as a satisfactory generic character. The former depends much on accident (I have specimens of typical *Berosus* before me with the head so declivous as to be quite vertical), and the latter is a mere question of degree. At the same time the insect has a very distinctive appearance, and does not seem at home in *Berosus*. Structurally I can see very little to distinguish it, but its hind tarsi much narrower on their widest face than their tibiæ, together with the dense pubescence of the elytra, and the transverse nature of the prothoracic sculpture are very noticeable characters. In the specimens before me the thorax is not *much* more rounded behind than that of some species of *Berosus*. I believe the genus to be a good one.

BEROSUS MAJUSCULUS, sp.nov.

Oblongo-ovatus; convexus; supra testaceus; palpis apice, capite postice, prothoracis disco, et elytris, plus minus fusco-nigro-notatis; subtus (capite antice, prosterno in parte, abdomine nonnullis exemplis postice, et pedibus, pallidis exceptis) niger; capite prothoraceque fortius minus crebre punctulatis; elytris apice emarginatis, punctulato-striatis, interstitiis planis subtilius nec crebre punctulatis. Long 2½-4 lines, lat. 1-1⅔ line.

Compared with the European *B. spinosus* this insect is more elongate (with the elytra at their widest very evidently behind the middle and much more elongated to a point behind), the puncturation of the head and thorax is a little finer and not nearly so close, and the striæ on the elytra are a little stronger.

The color is rather variable; the head is usually yellowish-brown, becoming darker behind, but in some examples the clypeus is pale lemon yellow and in others there is hardly any posterior infuscation; the prothorax is yellowish-brown, generally with an elongate dark vitta on either side of the middle; the elytra are very pale fuscous, clouded with a much darker tinge, except along the lateral margins and at the apex,—generally to such an extent that the ground colour is more or less overborne, the darker shade here and there forming rather distinct large blotches. The sculpture of the elytra is quite uniform, not becoming feebler either laterally or apically. The elytra are drawn out considerably at the apex, the apex itself being more or less strongly emarginate, the sides of the emargination being about equal and more or less sharply pointed. The underside is rugosely finely and very closely, but not deeply, punctured. A carina runs along each of the sterna; on the metasternum, however, it is very feeble and is cleft to form the sides of a small smooth central slit; the elevated flattened central space of the metasternum is very well defined and sharply pointed behind, its point projecting considerably between the hind coxæ. In the female the hind body of dried specimens is of very small size, its plane is very much below that of the metasternum not filling up a quarter of the space included in the cavity of the elytra, its apex bears two long testaceous filaments, its ventral segments are of even length (or nearly so) all across, and the antepenultimate is about the same width as the penultimate. In the male the hind body is much larger and has no apical filaments, its third and fourth ventral segments are much longer at the sides than in the middle, and the fifth is very much longer in the middle than the fourth, the hind margin of the fifth segment, moreover, being raised into a prominence on either side of the middle, and each of these prominences running backward on the segment as a scarcely defined carina, the intervening space being flattened. The base of the femora is sculptured as in the following species, but the part so sculptured being unicolorous with the rest of the surface, is less noticeable.

Two Australian species of *Berosus* with elytra apically emarginate have been previously described, *Australiae*, Muls., and *externespinosus*, Fairm. *B. majusculus* differs *inter alia* from the former of these by the striae and puncturation of its elytra being even over the whole surface (or perhaps slightly stronger near the apex), from the latter by the equality of the apical points of the elytra (which, however, may be a variable character) and by its unicolorous legs.

Widely distributed in South Australia; I have seen specimens from Port Lincoln, Adelaide, and Sedan.

B. GRAVIS, sp. nov.

Oblongo-ovatus; convexus; supra testaceus; palpis apice summo, capite postice prothoracis disco, et elytris, plus minus fusco-nigro-notatis ; subtus (capite, et prosterni lateribus, pallidis exceptis) piceus vel nigro-fuscus; capite antice sparsius subtilius postice gradatim crebrius fortius, prothorace fortius etiam sparsius, punctulatis; elytris apice emarginatis, punctulato-striatis, interstitiis planis subtilius nec crebre punctulatis ; pedibus testaceis femoribus 4 posterioribus basi nigris.

[Long. 3¾-4¼ lines, lat. 1⅗-2 lines.

This fine large species is closely allied to *B. majusculus* from which it scarcely differs in the color and markings of the upper surface. On the underside its entirely testaceous head and the four hinder femora nearly black in their basal two-thirds distinguish it. In respect of sculpture the front part of the head is much more finely punctured than in *B. majusculus*, the remainder of the sculpture presenting little distinction. The head is proportionately much narrower and more elongate. In the male the dilated joints of the anterior tarsi are much wider, and the surface and hind margin of the 5th ventral segment arc quite simple. In the females that I have seen the apical filaments are wanting, but they may have been accidentally broken off. From *B. Australiæ*, Muls., this insect may be known by its elytral sculpture not becoming feebler near the apex ; and from *B. externespinosus*, Fairm., by the two apical spines of each elytron being about equally developed.

In various localities in South Australia, Finniss River, Murray Bridge, &c.

B. DECIPIENS, sp.nov.

Oblongo-ovatus ; convexus ; supra testaceus, palpis apice vix infuscatis ; prothorace fusco-irrorato, antice vix infuscato ; elytris fusco-irroratis et maculatis ; subtus ferrugineus, femoribus ab-domineque fusco-notatis ; capite prothoraceque æqualiter sparsim sat subtiliter punctulatis ; elytris apice leviter emarginatis, punctulato-striatis, interstitiis planis sat fortiter punctulatis.

[Long. 3½ lines, lat. 1⅗ line.

Resembles *B. majusculus*, but with the puncturation of the head and prothorax very much finer and more sparing. The punctures are of a fuscous color, but otherwise those parts are almost unicolorous. The sculpture of the elytra is scarcely feebler near the apex than in front ; their apex is only minutely emarginate, with the sides of the emargination scarcely spiniform (probably a variable character), and their infuscation is very undefined, show-ing no tendency to be concentrated into a fascia. On the underside the punctured part of the femora is infuscate, and the segments of the hind body are transversely marked with blackish-brown. The specimen before me is a female, and presents no very conspicuous sexual character that I can find beyond the slenderness of the front tarsi. The fine sparing puncturation of the head and prothorax (a little finer on the clypeus but otherwise even in distribution and intensity) seems the most distinctive character of this insect amongst those of its Australian allies that have the elytra emarginate at the apex. From *B. Australiæ*, Muls., (in the description of which the puncturation of those parts is not mentioned), it seems to differ widely in color and markings, also in having (so far as the example before me is concerned) the elytra not distinctly spined at the apex, and striæ 4-6 of the elytra not differing from the rest.

Taken in the Northern Territory of South Australia by Mr. J. P. Tepper.

B. DUPLO-PUNCTATUS, sp.nov.

Ovatus; sat brevis; fortiter convexus; supra fuscus, capite et prothoracis disco æneis, cupreo vel aureo micantibus, elytris nigro punctulatis et maculatis; subtus niger, palpis (apice excepto), antennis, pedibus (femoribus 4 posticis basi exceptis), et prosterni lateribus, testaceis; capite prothoraceque rugose fortiter crebre (huic interstitiis subtiliter perspicue) punctulatis; elytris apice rotundatis, fortiter crenato-striatis, interstitiis sub-convexis subfortiter nec crebre punctulatis.

[Long. 2½-3 lines, lat. 1⅕-1⅔ line.

An extremely convex species; viewed from the side the elytra appear considerably more than half as high (*i.e.*, from the level of the lateral margin to that of the suture) as long. The blotches on each elytron are as follows : one on the shoulder, two down the suture almost touching it, and one near the lateral margin, but in some examples they are ill-defined, and in some examples some of them are wanting; the striæ and punctures on the elytra are blackish. Compared with the European *B. luridus*, Linn., this species is even more convex (especially about the hinder part of the elytra), its head and prothorax are more coarsely punctured, its scutellum is more elongate, and the elytral interstices are a little more convex. It differs from *B. luridus* also in the absence of any raised line on the prothorax, and in the presence on that segment of a system of very distinct (though small) punctures interspersed among the larger ones. The puncturation of the underside is close and fine, but rugose. The sternal keel is traceable only on the mesosternum (which is more declivous than in *B. luridus*), where, however, it is extremely sharply elevated, its hinder edge being truncated and standing out between the intermediate legs much above the level of their coxæ (a similar structure but less developed is seen in *B luridus*). The flattened space on the middle of the metasternum is somewhat trapezoidal, having its narrowest end directed backwards, but not at all passing the front margin of the hind coxæ, and bearing a large fovea in the centre. The basal third part of the lower face of the four posterior and (obsoletely of the anterior) femora (as in *B. luridus*) is nearly black and is opaque, densely and very finely punctulate, and minutely pubescent. My three specimens are of the same sex, apparently female; the 5th ventral segment is widely and somewhat squarely emarginate at the apex, each side of the emargination forming a strong spine; the 6th segment is barely discernible, projecting from the crenulated base of the emargination.

Probably allied to *B. ovipennis*, Fairm., (from Queensland), but much larger, without elevated or impressed lines on the head, with elytral interstices not flat, &c., &c.

Adelaide and Port Lincoln.

B. DISCOLOR, sp. nov.

Oblongo-ovatus; convexus; supra testaceus; capite et prothoracis macula postica nigro-viridibus, elytris fusco-maculatis; subtus niger; palpis (apice infuscato excepto), antennis, prosterni lateribus, et pedibus testaceis; capite rugulose subtilius confertim, prothorace fortius sparsius haud rugulose, punctulatis; elytris minus fortiter punctulato-striatis, interstitiis planis confuse sparsius punctulatis, apice leviter spinoso.

[Long. 2 lines, lat. 1 line (vix).

In my type of this insect the prothorax is yellowish testaceous, while the color of the elytra inclines to pale fuscous. The spot on the prothorax is transverse, and occupies the middle third part of the surface being about half as long as the whole segment, and placed just in front of (but not touching) the base. On each elytron there is an elongate fuscous spot on the first interstice at about a quarter of its length from the base, a larger and blacker one spreading out on the 2nd interstice a little less than half-way from the former to the apex, and a third close to the lateral margin at about half its length ; the punctures and striæ are dark fuscous or black. The puncturation of the head is rugulose but very fine and close (much more so than in *B. duplopunctatus*), that of the prothorax smooth and neither strong nor close (not unlike that of *B. majusculus*). The elytral sculpture resembles that of *B. majusculus*, except that the interstices are much more closely punctured ; the whole organs too are considerably less drawn out towards the apex than in that species, the apex itself being not emarginate but produced in a short sharp spine. The underside of the male closely resembles that of *B. majusculus*, except in having the whole undersurface of the head black or nearly so, and the hind margin of the metasternum less produced backward. The palpi are not narrowly tipped with black as they are in *B. majusculus*, but are rather strongly infuscate in the apical two-thirds of the last joint.

I possess a single example (also a male) from the same locality as that described above, which differs from the type in being slightly smaller, and of darker color, with a somewhat foveiform impression at the middle of the clypeal suture, and also in being somewhat more coarsely sculptured throughout, the elytral interstices especially being less flat and more strongly punctured ; as the specimen is not in very good condition and the differences are all rather slight, I abstain from bestowing a new name upon it, but I think it can hardly be regarded as a mere variety of *B. discolor*.

The unispinose apex of the elytra will distinguish *B. discolor* from all the hitherto described Australian species of *Berosus*, unless it be *B. sticticus*, Fairm., the elytra of which are stated to be "obtusely acuminate behind;" but even if this expression indicates a similarity in the apex of the elytra, the head of *B. sticticus* is said to be "almost impunctate in front" while that of *B. discolor* is punctured over its whole surface more closely than in any other *Berosus* known to me.

Port Lincoln.

B. FLINDERSI, sp.nov.

Oblongo-ovatus ; convexus ; supra testaceus, capite prothacisque disco æneis vel nigroæneis ; elytris fuscis nigro maculatis ; subtus niger, palpis (apice anguste nigro excepto), antennis, prothoracis lateribus et pedibus testaceis ; capite prothoraceque crebrius sat fortiter punctulatis ; elytris fortiter punctulato-striatis, interstitiis subconvexis fortius nec crebre punctulatis, apice sat acutis.

[Long. 1⅔ line, lat. ⅗ line.

I have examined many specimens of this insect without finding any tendency to vary in size. The brassy spot on the prothorax occupies the whole disc save that the anterior margin, or base, or both, may be narrowly testaceous. The black spots on the elytra are placed much as in *B. discolor*, but generally that near the lateral margin is alone conspicuous, the others being almost lost in a fuscous cloudiness that suffuses the entire disc. The head and prothorax are rather evenly and moderately strongly and closely punctured; the latter bears traces of the basal end of a longitudinal channel, and the former is a little foveated at the middle of the clypeal suture. The striæ on the elytra are nearly as deep as, and the interstices not much less convex than, those of *B. duplopunctatus*, the latter being punctured scarcely less strongly than the prothorax. In the male the fifth ventral segment is somewhat roundly truncate, in the female it is roundly emarginate in a somewhat upward direction, and two filaments (broken off in most examples) project from the apex of the very small sixth segment. In both sexes there appears to be a very minute triangular excision in the middle of the apical margin of the fifth segment.

The resemblance of this insect to the European *B. luridus* is extremely close. It is much smaller and somewhat more elongate, and less convex. In color and sculpture there is scarcely any difference except that the punctures in the elytral striæ are smaller and closer. The sexual characters, however, differ considerably, and the mesosternal carina is of somewhat even prominence, not raised up at the hinder apex (as it is in *B. luridus*) into a free erect process. The femora are sculptured at their base as in *B. duplopunctatus*.

The species of *Berosus* described by M. Fairemaire, are all said to have their elytra either "emarginate," or "obtuse" at the apex, with one exception,—*B. sticticus*,—in which however the front of the head is said to be impunctate ("lisse").

Not rare near Port Lincoln.

B. SIMULANS, sp. nov.

Oblongo-ovatus; convexus; supra testaceus; capite prothoracisque macula nigro-æneis, elytris vix fusco-maculatis; subtus niger, antennis palpis pedibusque testaceis : capite crebre, prothorace sparsius, fortiter punctulatis; elytris sat fortiter striatis, striis punctis magnis nec profundis instructis, interstitiis vix convexis fortiter nec crebre punctulatis, apicibus separatim spinoso-acuminatis. Long. 1½ line, lat. ¾ line (vix).

Resembles the preceding, but the head is more, and the prothorax less, closely punctured. The punctures in the elytral striæ are long and shallow. In the example before me the elytra are separately pointed in the form of a small acute spine, but the value of this character is doubtful, as I find some variety in the form of the apex of the elytra in most species of *Berosus* of which I have seen numerous specimens. In one specimen of *B. Flindersi* the apex of one elytron is much more acute than that of the other. The example before me is a male; it is pinned in such fashion that I cannot bring a strong lens to bear on its ventral segments, but as far as I can see the 5th ventral segment is gently emarginate, the middle of the emargination being thickened, and its sides slightly pointed. The palpi are entirely testaceous, while in *B. Flindersi* they are conspicuously infuscate at the apex. The dark mark on the prothorax is not very sharply defined and occupies the hinder half of the disc—but is probably variable.

In the Adelaide University Museum; taken in fresh water near Rivoli Bay.

HYDROCHUS ADELAIDÆ, sp. nov.

Sat elongatus; brunneus, capite prothoraceque obscurioribus, palpis (apice excepto) antennis pedibusque testaceis, tarsorum apice nigro ; capite prothoraceque fortiter rugulose punctulatis ; hoc vix inæquali ; elytris crasse profunde seriatim punctulatis, interstitiis vix conspicuis.   Long. 2 lines (vix), lat. ½ line.

The clypeus is nearly black, shining and not very closely, strongly or roughly punctured, the clypeal suture fairly defined and a little angulous behind, the hinder part of the head obscurely and very roughly punctured.   The prothorax is punctured as the hinder part of the head ; it is slightly wider than long, nearly twice as wide in front as behind, its sides are gently rounded in the front two-thirds, nearly straight behind, and owing to the coarseness of the sculpture they appear to be strongly crenulate ; the surface bears a vague elongate impression on either side near the posterior angles with an obscure rounded impression close in front of its anterior end.   The elytra are at their widest behind the middle ; their shoulders are quite rounded off ; their sculpture consists of ten rows of round deep foveæ, the foveæ near the base, apex and lateral margins smaller than the rest ; the rows are not very regular and the intervals between them are not longitudinally continuous, being in width and prominence similar to those separating fovea from fovea in the rows ; the sutural line is very elevated.

I know no other species closely resembling this, but in order to make a comparison I may say that, placed beside the European *H. angustatus*, Germ., independently of color, the prothorax is wider in front and shorter, and the elytra are much wider behind with very much less regular and coarser puncturation.

I have a single specimen taken in the River Torrens near Adelaide.

H. REGULARIS, sp.nov.

Elongatus ; brunneus, capite prothoraceque obscurioribus, antennis pedibusque brunneo-testaceis, tarsorum apice piceis ; capite prothoraceque fortiter rugulose punctulatis ; hoc inæquali ; elytris fortiter seriatim punctulatis, interstitiis manifeste notatis.

[Long. 1⅔ lines, lat. ⅔ line.

The whole of the head is uniformly punctured, and there is a well-defined narrow longitudinal furrow between the eyes. The prothorax is much more coarsely and less closely punctured than the head ; it is decidedly transverse and not very much wider in front than behind ; its sides are only slightly rounded in front, scarcely sinuate behind, and with very little appearance of crenulation ; its surface is uneven in a manner very difficult to define, —from the centre of the base an obscure rib runs forward about a third the length of the segment, and then forks into two branches which run obliquely for a short distance towards the anterior angles and then turn towards the front margin and fade into the general surface, from the point where they turn forward a similar rib being given off on either side which runs back in a slight curve nearly to the base half-way between the central rib and the posterior angles ; the spaces among and immediately outside all these ribs are obscurely depressed. The elytra are very slightly at their widest behind the middle ; their anterior margin is very strongly defined, their suture strongly elevated ; each of them bears ten very regular rows of square, and deep, but not very large punctures, the intervals between the rows being well defined, the intervals outside the middle a little the most conspicuous.

Compared with the preceding this is a narrower and more parallel species, with defined inequalities on the prothorax, and elytra very much more finely punctured, the punctures in even rows and the interstices between the rows regular. The elytral sculpture is not unlike that of *H. angustatus*, Germ., but the interstices between the rows are narrower and more defined.

I have a specimen of this insect from Murray Bridge, and another from Western Victoria.

H. Victoriæ, sp.nov.

Sat elongatus ; piceus, antennis, palpis pedibusque dilutioribus, tarsorum apice nigris ; capite prothoraceque fortiter rugulose punctulatis ; hoc vix inæquali ; elytris fortiter confuse sculpturatis, interstitiis alternis sat fortiter elevatis.

[Long. 1⅔ lines, lat. ⅔ line.

The head and prothorax are coarsely and very rugosely sculptured, the clypeal suture fairly defined, the prothorax with an ill-defined impression across the middle, which widens into a vague fovea on either side. The prothorax is equally long and wide, its front margin nearly twice as wide as its base, its sides strongly rounded, their edges (owing to the coarseness of the sculpture) appearing crenulate. The elytra are considerably dilated to behind the middle; their puncturation is very ill-defined, owing to the roughness and irregular elevation of the interstices between puncture and puncture in the rows which almost obliterate the interstices between the 1st and 2nd, 3rd and 4th, &c., rows; the interstices between the 2nd and 3rd, 4th and 5th, &c., rows are well-defined and keel-like in front, but become fainter behind, that between the 4th and 5th being the strongest and most continuous; the sutural line is well elevated.

A very distinct species. The punctures in the rows on the elytra are about the same size as in *H. regularis*, but are very much less distinct, the spaces between the raised alternate interstices appearing irregularly transversely ribbed (or from some points of view almost tuberculate) rather than distinctly biseriately punctulate.

I have two specimens taken at Ararat, Victoria.

OCHTHEBIUS AUSTRALIS, sp.nov.

Sat latus; nitidus; antennis, palpis, pedibus et prothoracis marginibus plus minus testaceis; capite et prothorace fortiter inæqualibus vix perspicue punctulatis; elytris fortiter æqualiter punctulato-striatis.   Long. $\frac{4}{5}$-1 line, lat. $\frac{3}{10}$-$\frac{2}{3}$ line.

The clypeal suture is arched and very deep; the portion of the head behind it is sculptured in such fashion that there appear to be (when looked at from directly in front) a raised space at each corner and another in the centre, into which the surface is divided by extremely deep channels. The prothorax is decidedly transverse, of equal width in front and behind, with srtongly explanate

54

margins, the margin on either side of nearly even width and scarcely narrower than half the width of the disc; the middle of the disc is occupied by a very strong dorsal channel, on either side of which an equally strong impression runs from close to its base obliquely forward to about the middle of the lateral ridge of the disc; there is also a strong foveiform impression on either side just within the anterior corner of the disc; the surface of the explanate margins is noticeably but less definedly uneven; their hinder third part (or rather more) is membranous; their outline (inclusive of that of the membrane) is gently and evenly convex; the anterior angles are well-marked, sharp, and minutely pointed forward; the hinder angles of the membrane are scarcely defined. The surface of the elytra is not quite even and is marked with well-defined coarsely punctate striæ, which are scarcely enfeebled behind; the interstices are moderately convex and impunctate, the apex of the elytra is moderately pointed. The underside is rather shining, and I cannot find that it bears any distinct puncturation. I do not observe any notable sexual differences in the structure of the ventral segments.

No Australian species of this genus has been described hitherto, so far as I know. Of European species *O. bicolon*, Germ., is the one that *O. Australis* seems more particularly to resemble. Compared with it (apart from the difference in size), the maxillary palpi of *O. Australis* are much shorter and less stout, the inequalities on its head and prothorax are more strongly defined, its prothorax is not (or scarcely) punctulate, the sculpture of the elytra continues, without becoming obsolete, to the apex.

I have taken this little insect in the River Tod near Port Lincoln and in the Torrens near Adelaide.

Hydræna Torrensi, sp. nov.

Oblonga, postice minus dilatata; supra obscure livida, capite piceo; clypeo subtiliter, capite postice prothoraceque sat fortiter, punctulatis; hoc transversim et longitudinaliter biimpresso;

elytris subtiliter striato-punctulatis (striis, nec puncturis, postice
obsoletis) apice separatim acuminatis ; subtus nigra, antennis
palpis pedibusque lividis.   Long. 1 line, lat. ⅖ line.

The difference between the puncturation of the clypeus and of
the rest of the head is very noticeable. There is an obscure
longitudinal depression on either side between the eyes. The
head behind the clypeal suture is punctured very similarly to the
prothorax, rather coarsely and deeply but not very closely. The
prothorax is decidedly transverse by measurement, but to the eye
only very slightly so, its front margin decidedly narrower than its
base ; its puncturation (especially when viewed obliquely from
the side) seems to run in longitudinal wrinkles ; two transverse
depressions run from one lateral margin to the other, dividing the
surface into three nearly equal spaces ; a transverse depression
also runs obliquely from just within the posterior angle on either
side to the front margin, so that the entire surface is divided by
these four depressions into nine spaces, of which the three down
the middle are much larger than the rest ; the sides are moderately
arched, their greatest divergence from each other being behind
the middle, where they are angulated rather than regularly
rounded ; the anterior corners of the prothorax are rectangular,
and not at all produced forward ; the basal angles are similar, but
perhaps a trifle sharper. The elytra are finely and closely, but
very distinctly, punctulate-striate, and furnished with fine and
obscure pubescence which runs in rows along the striæ ; the
interstices are flat or nearly so, and do not show any defined
puncturation under a Coddington lens ; the striæ fail near the
apex, but the puncturation continues of even intensity, though it
becomes somewhat confused.

The underside is opaque and minutely coriaceous, the hind body
covered with minute obscure pubescence. If I have both sexes
the sexual differences are slight. The basal four ventral segments
are short and equal, the fifth much longer. The latter is
traversed by a very fine arched keel which commences on either
side of the apex, and runs backward in a curve so as almost to

touch the apex of the fourth segment in the middle. In what I regard as the female there is a very small sixth segment bearing two short apical setæ; in the other sex (as I take it to be) this segment is a little larger and without apical setæ, but these differences may be accidental.

This species appears to be somewhat variable; I possess specimens differing from the type described in having the puncturation of the prothorax considerably obliterated on the middle of the disc, and the angulate appearance of the sides of the same segment only feeble.

Probably allied to the Queensland *H. acutipennis*, Fairm., but differing from it considerably in the sculpture of the prothorax, in having the part of the head behind the clypeal suture punctured similarly to the prothorax, and in having the elytral striæ not "scarcely growing obsolete at the apex," but altogether disappearing before the apex. From *H. luridipennis*, Macl., it seems to differ in having the thorax transverse with its angles not acute.

Placed beside the European *H. angustata*, Sturm, the general form is shorter and broader, the prothorax is evidently more transverse with its surface rendered much more uneven by transverse and longitudinal depressions, the elytral striæ much more obsolete behind, and the elytra separately pointed instead of being rounded off. The puncturation does not differ much.

Near Adelaide; in the river Torrens.

N.B.—In the South Australian Museum there is a specimen of *Hydræna*, taken in Victoria by Mr. Tepper, which I refer with some doubt to this species. The elytra appear a little more strongly striated in front, and at the apex are much less acute, being separately *rounded* rather than distinctly acuminate. The hind tibiæ, moreover, are bisinuate on their inner edge; this latter may be merely a sexual character. In all other respects the two insects seem to be quite identical.

Volvulus punctatus sp.nov.*

Elliptico-ovalis ; sat nitidus ; niger ; antennis, palpis et pedibus anticis testaceis ; pedibus 4 posterioribus et abdominis lateribus plus minus rufescentibus; capite subtilius minus crebre, prothorace subtilius (disco sparsim, lateribus crebre), elytris fortius (disco sparsim, lateribus crebre) punctulatis ; elytrorum margine leviter bisinuato. Long. 2¼ lines, lat. 1¼ lines.

Compared with *V. scaphiformis*, Fairm., (I have little doubt that my specimen is correctly referred to it ; it is certainly at least very closely allied) the color of the present insect has no metallic tinge, the shape is very much wider and more rounded on the sides, the elytra are without any trace of the striæ except a sutural stria in the hinder half, and the edge of the elytra is lightly bisinuate instead of being gently concave along the whole length.

There are several specimens in the South Australian Museum, but without any record of particulars of capture.

Cyclonotum Australe, sp.nov.

Late ovatum ; nigrum ; antennis, palpis, tarsisque piceis vel rufescentibus ; creberrime punctulatum ; elytris striâ suturali impressa. Long. 3 lines (vix), lat. 1⅓ lines.

The resemblance of this species to the common European *C. orbiculare*, Fab., is so close that it would seem to be sufficiently described by the statement that it is much larger than that insect, with the palpi of a paler, and the legs of a darker, color, and the puncturation of the elytra slightly finer. Mr. Macleay has done me the favor of comparing it with his *C. Mastersi*, and tells me that it is a larger insect than his, and that its puncturation is coarser.

---

* In some respects (e.g. the very deep insertion of the prothorax in the front emargination of the elytra) this insect seems referable to *Globaria*, but it has not the pencil of long hairs at the apex of the hind tibiæ which are said to be characteristic of that genus.

Probably widely distributed in South Australia, but apparently not common. I have taken it near Adelaide, and it has been taken near Port Lincoln by Mr. J. Anderson.

CERCYON FOSSUM, sp.nov.

Breviter ovale; convexum; nitidum; piceum; antennis palpis pedibusque rufescentibus; capite prothoraceque sparsim subtiliter punctulatis; elytris striatis; striis crasse fortiter nec crebre, interstitiis subplanis sparsim subtilius, punctulatis.

[Long. 1 line, lat. ⅔ line.

This species does not very closely resemble any European *Cercyon* known to me, owing to the very coarse wide-set punctures in its elytral striæ. Placed beside *C. flavipes*, Fab., it is seen to be wider in proportion to its length (especially behind), with the puncturation of the head and prothorax very much more sparing and less noticeable; the punctures in the striæ on the elytra look large enough to allow the thin end of the claw-joint of the tarsi to be inserted into them, and are placed in the striæ with a well-defined interval between puncture and puncture. The interstices between the striæ are almost quite flat in front but become evidently convex near the apex; they are punctured sparingly but not very finely. The general surface of the underside is opaque, very finely punctulate, the metasternum being nitid and coarsely and sparingly punctulate. The sparing puncturation of the head and prothorax will in itself distinguish this insect from *E. dorsale*, Er. The tarsi are stouter, the basal joint of those of the hind legs shorter, than usual in the genus.

S. Australia.

CUCUJIDÆ.

LÆMOPHLŒUS DIFFICILIS, sp.nov.

Planus; nitidus; vix pubescens; testaceus; fronte æquali; prothorace vix transverso, utrinque fortiter striato, angulis anticis dentatis.   Long. 1 line, lat. $\frac{4}{15}$ line.

The head behind the clypeal suture is quite smooth without any impressions whatever, and is sparingly and moderately strongly punctured. The prothorax is equally long and wide, at its widest across the front margin where its angles are dentate, and thence contracted with a very slight curve to the base, the anglet at which are obtuse ; the fovea on either side of the disc is wide and strong, especially behind ; the surface between the foveæ is very flat and is finely and sparingly punctured ; the space outside the fovea is rather strongly declivous, and is punctured more strongly than the disc. The sculpture of the elytra seems to consist of two costæ (the innermost very narrow and obscure), which are about the 3rd and 5th interstices among a series of very obsolete punctured striæ. The scutellum is rather strongly transverse. The antennæ in the male are a little longer than in the female, about equal to the length of the elytra ; the basal joint rather long and stout (very evidently longer than wide), the 2nd narrower and considerably shorter, but wider than the remaining joints of which the next six are equal to each other in length and thickness, the apical three longer but scarcely thicker.

Allied to *L. testaceus*, Fab., but differing *inter alia* in the perfectly even surface of the head behind the clypeal suture, and the sparing puncturation of the prothorax. A few specimens under bark of a felled *Eucalyptus*, about 30 miles north of Port Lincoln.

L. Lindi, sp.nov.

Minus planatus ; sat nitidus ; vix pubescens; testaceus; fronte subtiliter canaliculata ; prothorace vix transverso utrinque subtiliter bistriato, sat crebre punctulato ; angulis anticis obtusis.

[Long. $\frac{9}{10}$ line, lat. $\frac{4}{15}$ line (vix).

The puncturation of the head is faint and not very easy to see clearly, but it is moderately close and has a tendency to run in longitudinal wrinkles ; the longitudinal furrow on the forehead is moderately well-defined and does not reach the clypeus or the back of the head. The prothorax is evenly (though of course not strongly) convex, and is about as long as wide; its front angles are quite obtuse, the basal ones rather sharp ; the base is nearly as

wide as the front margin, the sides being very gently arched; the
striæ on either side of the disc are extremely fine and very close
together, the external one scarcely continuously traceable; the
punctures on the disc are fine and moderately close, with a little
tendency to a longitudinal serial arrangement. The sculpture of
the elytra scarcely differs from that of the preceding species
except that the 7th interstice is costiform, whereas in *L. difficilis*
it is scarcely so. The scutellum is rather strongly transverse.
The antennæ in both the specimens before me are very stout and
reach back a little beyond the base of the prothorax; their joints
differ *inter se* very little in respect of width, the 1st being a little
wider than the 2nd, and the second slightly wider than the next
6, the 9th and 10th a little widest of all, the 11th rather elongate.

Allied to *L. bistriatus*, Grouvelle, but differing in the much
closer puncturation of the prothorax, the grooved forehead, &c., &c.

LAMELLICORNES.

BOLBOCERAS KIRBII, Westw.

In Mr. Masters' "Catalogue of Australian Coleoptera" this
appears as a distinct species. Prof. Westwood, however, in a
note to his description of it states that after his description was
in type he became possessed of evidence that it is only a variety
of *B. proboscideum*, Schreibers,—and I believe his opinion has
not since been controverted.

B. TATEI, sp.nov.

Nigro-piceum; clypeo antice leviter rotundato, fronte cornu
elongato erecto simplici, antice basi carina fortiter rotundata
instructo; prothorace lateribus et in foveolis lateralibus rugoso,
præterea lævi, retuso, utrinque cornu brevi acuminato porrecto;
parte retusa permagna profunde excavata sparsius nec fortiter
punctulata, intus marginem versus aureo-hirsuta; elytris leviter
punctulato-striatis; tibiis anticis 5 dentatis ♂.

[Long. 8½ lines, lat. 4⅓ lines.

In the example before me of this very fine insect the frontal horn rises to about the level of the top of the prothorax, and is quadrangular in shape at the base, a keel running from each of the front angles about a quarter of the distance up it, and from each of the hind angles about two-thirds the distance to its summit ; the upper part is cylindric and tapering.   The prothorax is strongly and roughly punctured on the sides, this sculpture being continued up the portion adjacent to the excavation, but the middle and basal parts are smooth, except in the deep basal furrow where, however, the puncturation is wanting in the middle ; the excavation occupies the larger part of the whole surface, and is cavernous, with sharply defined limits especially on the sides; the horns (or teeth) are about half as long as the distance from the base of the prothorax to that of the excavation, are compressed, and triangular, their wide face being about as wide across the base as the lower (which is the longer) side of their outline ; they are situated on either side of the excavation, about half-way down the declivous face of the prothorax and project forward, and slightly upward ; the portion clothed with golden hairs is the inner surface of the overhanging margins.

The nearest ally of this species is, I think, *B. cavicolle*, Macl., from which it differs *inter alia* by the transverse carina in front of its frontal horn being evenly arched forward and not at all turned up in the middle, by the absence of a tooth on the hind surface of the frontal horn, by the very much larger excavation on its prothorax, by the presence of golden pubescence within the same, by the much more feebly punctulate striæ of its elytra, and its very much more slender front tibiæ, the inner margin of which is evenly and gently concave from the base to the apical spine.

Northern Territory of S. Australia ; taken by Professor Tate.

B. GLOBULIFORME, Macl.

The description of this species does not indicate any satisfactory distinction from that of *B. rotundatum*, Hope.   There are specimens before me evidently, I think, appertaining to the latter species and taken by Mr. J. P. Tepper, at Port Darwin, which I cannot distinguish from a short series in the S. Australian Museum ticketed as having been taken at Rockhampton, Queensland (*B. globuliforme* was described on specimens from Port Denison, Queensland), and which agree very well with the description of either insect. If *B. globuliforme* is a good species it would seem desirable for it to be re-described and its distinction from *B. rotundatum* pointed out. A study of the descriptions of *B. rotundatum* and *rubescens* by both Hope and Westwood, together with the figures supplied by the latter, creates a doubt, moreover, of the distinctness of the two; the figure shows a difference in the shape of the prothorax that would appear satisfactory enough, but unfortunately it is not alluded to by either author in *stating* the differences between them.

B. SIMPLICICEPS, sp.nov.

Rotundatum, rufum, supra glabrum ; clypeo in medio in tuberculum elevato, lineis 4 elevatis ex hoc tuberculo prodeuntibus, scil. 2 ad angulos clypei anticos, 2 (postice tuberculatim elevatis) ad basin antennarum, inter has lineas interveniis fortiter rugose punctulatis; capite postice æqualiter leviter convexo, sparsim minus fortiter punctulato; prothorace ad latera fortiter acervatim punctulato, fossulato, vix evidenter canaliculato (canali obsoleto punctis nonnullis indicato) antice breviter retuso, spatio retuso longitudinaliter 3-sulcato, inter sulcos interspatiis postice subtuberculatis; elytris leviter 9-striatis striis subtiliter punctulatis; tibiis anticis externe 5-dentatis.   Long. 4 lines, lat 2⅔ lines.

The nine striæ of the elytra do not include the one close to the lateral margin, which bends inward from the margin near its base like an oblique fovea ; of the nine the nearest two to the suture are much stronger than the others. In the five specimens that I have seen there is very little difference likely to be sexual ; in one of them the lateral two of the three longitudinal furrows on the retuse part of the prothorax are almost obsolete, and the puncturation of the head differs, the rather large punctures of the front and back of the vertex not failing altogether in the intermediate part, whereas in other examples the middle of the disc (from about the level of the middle of the eye hindward almost to the base) is quite devoid of such punctures, having only a few exceedingly fine ones.

This species must be very like *B. planiceps*, Macl., (from Sweer's Island). If I understand the description of that species rightly, however, the thorax is differently sculptured in front; and in any case the front tibiæ are very different, having five well-defined external teeth increasing in size from the topmost downward, without any trace whatever of a sixth, while *planiceps* is described as having six external teeth on the front tibiæ, of which the basal two are subobsolete.

Northern Territory of South Australia; collected by Professor Tate.

B. FENESTRATUM, sp.nov.

Rotundatum; rufum vel piceum; supra glabrum; clypeo in medio vix in tuberculum elevato, lineis 4 ex hoc quasi tuberculo prodeuntibus, scil. 2 ad angulos clypei anticos, 2 ad basin antennarum; capite toto dense rugose crasse punctato, vertice medio carina transversa arcuata instructo; prothorace ad latera fortiter punctulato, fossulato, vix evidenter canaliculato (canali obsoleto punctis quibusdam magnis indicato); antice vix retuso, parte antica fovea magna subquadrata punctulata instructa; elytris usitate striatis, striis profundis leviter vel vix evidenter punctulatis; tibiis anticis externe 6 vel 7-dentatis, dentibus apicalibus 4 solis distinctis. Long. 4 lines (vix), lat. 2⅗ lines.

The fovea on the front declivous part of the prothorax is of peculiar form, being nearly square on the surface but ending almost in a point at the bottom as though a quadrangular pyramid had been cut out; in some specimens this is very sharply defined, in others less so, but in all that I have seen some (at least) of the declivities of the excavation are sharply triangular. In the specimen I have described the front tibiæ have the apical

four teeth well defined though blunt, and above them the external
edge is cut on one tibia into two, on the other into three, obsolete
teeth; in several examples I find all the lower teeth decidedly
sharper than in the type, and the upper not quite so feebly
developed. If I have both sexes of this insect before me, the
sexual distinctions are very slight, but in that case probably the
specimen described is a female, as I have a much mutilated
example in which the clypeal tubercle is better developed than
usual, with the lines running from it to the bases of the antennæ
evidently more elevated.

Northern Territory of South Australia; collected by Dr. Wood,
also by Professor Tate.

MÆCHIDIUS CAVICEPS, sp nov.

Oblongus; sat convexus; minus nitidus; rufo-piceus; setosus;
capite prothoraceque fortiter crasse, elytris (seriatim, seriebus
haud geminatis) subtilius, punctulatis; clypeo antice sat longe
producto, utrinque fortiter concavo, antice triangulariter emarginato; unguibus haud simplicibus.

[Long. 4 lines, lat. 2 lines (vix).

The clypeus is considerably produced almost at a right angle
with the rest of the head; its emargination is sharply triangular,
the sides of the same being acutely pointed in front; the deep
concavity on either side is extremely shining. The prothorax is
nearly twice as wide as down the middle it is long; its sides
(owing to the roughness of the surface sculpture) are strongly
crenulate, strongly (almost angularly) rounded behind the middle
and not at all sinuate; its hind angles are slightly obtuse, but
very nearly right angles; the granules of the punctures on the
surface fill them up and protrude above them, making the
prothorax appear almost tuberculate. The elytra are a little
dilated behind, where they are nearly a third as wide again as
the prothorax; they are punctured in regular rows which have
no tendency to run in pairs, the punctures in the rows being
decidedly small as compared with those of the generality of species

of *Mœchidius*; the interstices are quite flat, impunctate, and rather nitid. The hind femora are not particularly stout, the basal joint of the hind tarsi is not much longer than the second, and the claws have the quill-like appendage frequent in species of this genus.

Seems to resemble *M. bilobiceps*, Fairm., and *Albertisi*, Fairm., but both those species are described as having the rows of punctures on the elytra running in pairs, the former having the thoracic sculpture obsolete at the sides, the latter the sides of the prothorax sinuate behind, besides other differences. From *M. rufus*, Hope (of which I have a specimen before me), the totally different structure of the hind legs, besides many other differences, will at once distinguish this insect.

Northern Territory of S. Australia; taken by Mr. J. P. Tepper.

LIPARETRUS PALMERSTONI, sp.nov.

Ovatus; sat nitidus; ferrugineus, capite postice et (nonnullis exemplis) tibiis, tarsis, abdomineque obscurioribus; longe sparsim hirsutus, sat sparsim punctulatus; tarsorum posticorum articulo primo secundo breviori; tibiis anticis externe haud dentatis; clypeo concavo producto,—maris antice late emarginato, angulis anticis subacutis, – feminæ antice truncato, angulis anticis rotundato-obtusis; antennis 9-articulatis.

[Long. $2\frac{1}{3}$ lines, lat. $1\frac{1}{3}$ lines (vix).

This species belongs to Mr. Macleay's first section of the genus, and is much smaller than any other of that section previously described. The long sparing pubescence with which it is clothed seems to be entirely wanting on the elytra, which are very short, leaving a great deal of the propygidium exposed,—neither it nor the pygidium has any trace of a keel. The inner two pairs of striæ on the elytra are tolerably distinct, their interstices however being punctured in a coarse rather sparing manner uniformly with the general surface. The hind tibiæ and tarsi are nearly black in most examples. The anterior tibiæ are much prolonged at the apex.

Northern Territory of S. Australia; collected by Mr. J. P. Tepper.

L. POSTICALIS, sp.nov.

Ovatus; minus nitidus; ferrugineus vel ferrugineo-piceus, capite nigricanti; minus hirsutus; capite confertim rugose, prothorace fortius vix crebre, elytris fortiter sat sparsim, pygidio haud evidenter, punctulatis; elytris sat evidenter geminato-striatis; clypeo antice rotundato; tarsorum posticorum articulo primo secundo vix breviori; tibiis anticis externe obtuse 3-dentatis; antennis 9-articulatis.   Long. 3⅔ lines, lat. 2 lines (vix).

The specimens before me may be slightly abraded; their upper surface is glabrous excepting a fringe of longish stout hairs on the sides of the prothorax, but probably in fresh examples the pygidium and propygidium are thinly clothed with long hairs. The elytra have three very distinct pairs of punctate striæ, and a fourth much fainter, the punctures in the striæ being strong and rather close, but not different in character from those with which the interstices between the pairs are rather sparingly sprinkled. The puncturation of head, prothorax, and elytra is successively more and more sparing. The pygidium and propygidium are finely coriaceous and almost opaque; under a strong lens indications of scarcely impressed and very sparing punctures may be traced, especially towards the apex of the latter; there is no trace of a keel. The prothorax is scarcely channelled longitudinally. The elytra are about twice as long as the prothorax down its middle line. The underside is thinly clothed with long hairs. This species may be distinguished from *L. picipennis*, Germ., and *L. atriceps*, Macl.,—both of which it resembles in many respects,—by its opaque impunctate pygidium, and from the latter by its strongly punctate elytra. The specimens before me are probably females.

Northern Territory of S. Australia; collected by Mr. J. P. Tepper.

N.B.—An example of *Liparetrus* before me from the same locality as the above, differs in having the elytra considerably shorter; I hesitate to consider it a distinct species, as I can discover no other distinction; possibly it is the other sex.

L. JUVENIS, sp.nov.

Ovatus; sat nitidus; ferrugineus; capite toto, mesosterno et metasterno (his parte media excepta), piceis; minus hirsutus; capite fortius, prothorace subtilius, elytris fortius, pygidio propygidioque leviter, sparsim punctulatis; elytris sat evidenter geminato-striatis; clypeo antice rotundato; tarsorum posticorum articulo primo secundo vix breviori; tibiis anticis externe fortiter 3-dentatis; antennis 9-articulatis.   Long. $3\frac{1}{2}$ lines, lat. $1\frac{1}{3}$ lines.

Closely allied to L. *posticalis*. The prothorax is a little more transverse, being quite twice as wide as it is long down the middle, otherwise there seems to be no difference in respect of general form, proportions, or distribution of hairs. The head is quite smoothly and sparingly punctulate,—the prothorax and elytra are more finely punctulate than those of L. *posticalis*; the latter scarcely differ otherwise. The puncturation of the pygidium and propygidium (which have no trace of a keel) is very distinct, though sparing and lightly impressed. The lower two teeth of the anterior tibiæ are very strong and sharp. The prothoracic channel is scarcely traceable.

Like L. *posticalis* this species belongs to the *picipennis* group of *Liparetrus*. The sparing puncturation of its head distinguishes it from most of the members of that group, its color and structure of the legs from nearly all the remainder. Judging by the very brief description of L. *latiusculus*, Macl., it probably resembles that insect, but is much larger, and apparently of a very different color, and no doubt differs in other respects.

Northern Territory of S. Australia; collected by Mr. J. P. Tepper.

L. FALLAX, sp.nov.

Ovatus; sat nitidus; minus hirsutus; ferrugineus, capite (nonnullis exemplis) piceo; hoc confertim rugose, prothorace fortius nec crebre, elytris crebrius sat fortiter, pygidio crebrius leviter, punctulatis; elytris sat evidenter geminato-striatis, clypeo antice et ad latera truncato, tarsorum posticorum articulo primo secundo paullo breviori; tibiis anticis externe obtuse 3-dentatis; antennis 8-articulatis. Long. 4 lines, lat. 2 lines.

In this species the clypeus has the form (although not very sharply defined), which Mr. Macleay, in his monograph of the genus, calls " presenting three truncate faces." Its general resemblance to *L. posticalis* is considerable. The head is punctured almost exactly as in that species; the prothorax is more transverse, being quite twice as wide as it is long down the middle, and is slightly more sparingly and finely punctured. The elytra are a little longer in proportion, being more than twice as long as the prothorax, with slightly finer and closer puncturation, and the striæ of each pair a little closer to each other. The tarsi are evidently more slender. No species hitherto described is very close to this.

Northern Territory of S. Australia; collected by Mr. J. P. Tepper.

Lepidiota Darwini, sp.nov.

Ferruginea, supra sat sparsim, subtus densissime, albo-squamulata; capite lato minus convexo, clypeo perbrevi antice in medio reflexo emarginato; prothorace fortiter convexo, sat transverso, antice angustato, lateribus post medium rotundato-ampliatis postice fortiter sinuatis, angulis posticis acutis; tibiis anticis tridentatis.

[Long. 10½-12 lines.

Mas (?) Angustus, elytris subparallelis.

Fem. (?) Latior, elytris postice ampliatis.

The sides of the clypeus, and the head backward to the level of the middle of the eyes, are coarsely and sparingly punctured, each puncture being filled up with a white scale (which, however, does not, or not much, protrude beyond it); the length of the clypeus down the middle line is much less than half the length of the coarsely punctured part behind it of the head; the clypeal suture is regularly and widely bisinuate; the part of the head behind the coarsely punctured portion is smooth except that there are some very fine (in the examples before me scaleless) punctures in the middle. The prothorax at its widest part is slightly more than half again as wide as it is long down the middle ; its base (which is bisinuate) is about half again as wide as its front margin (which is decidedly though gently emarginate) ; its front angles are obtuse, but well defined, those at the base pointed and a little directed outward ; its lateral margins are slightly crenulate and diverge in nearly straight lines to behind the middle where they are strongly and suddenly rounded, thence proceeding with a rather strong sinuation to the base ; its surface is punctured and scaled uniformly with the coarsely punctured part of the head, although a little more closely towards all the margins than on the disc ; there is a broad thickened margin all across the front, which is punctured and scaled uniformly with the neighbouring surface. The elytra and scutellum are punctured and scaled very similarly to the prothorax, and the former bear obscure indications of three or four wide scarcely convex costæ. The propygidium, pygidium and the entire undersurface are densely covered with closely packed white scales which entirely conceal the derm, and the sterna are clothed rather thickly with long white hairs. The legs are coarsely and sparingly punctured and scaled ; the front tibiæ are tri-dentate externally, the upper tooth very small.

The above description is founded on the only really fresh specimen before me,—which I take to be a female. Its elytra are considerably dilated backward to near the apex and the apical ventral segment is more than half the length of the penultinate with its apical margin evenly rotundate-truncate. In what I take to be the male the head and prothorax are of a pitchy color (probably merely an individual variety), the surface of the prothorax is obscurely uneven through the presence of some irregular ill-defined scarcely convex ridges, the elytra are extremely parallel, the apical ventral segment is much less than half the length of the penultimate with its apical margin widely and feebly bisinuate, and the whole insect is much narrower than the other sex. The antennæ of these two specimens do not differ noticeably in structure.

55

Northern Territory of S. Australia; collected by Prof. Tate and Mr. J. P. Tepper.

L. DELICATULA, sp.nov.

♀. (?) Ferruginea, supra confertim subtiliter, subtus densissime, albo-squamulata; capite minus lato minus convexo; clypeo minus brevi, antice in medio sat fortiter reflexo-emarginato, ad latera fortiter rotundato; prothorace sat convexo, sat transverso, antice angustato, lateribus post medium ampliato-rotundatis postice haud sinuatis, angulis posticis obtusis; tibiis anticis fortiter tridentatis; elytris postice fortiter ampliatis.　　　　　　　　Long. 10 lines.

Sexus alter latet.

The clypeus and head (backward to the level of the middle of the eyes) are uniformly sparingly and rather strongly punctured and scaled; the length of the clypeus down its middle line is not much less than that of the coarsely punctured part behind it of the head, the sides of the same being quite strongly rounded; the clypeal suture is unevenly bisinuate, the middle part of the sinuation (with its convex side running up the head) being very much more strongly curved than the lateral (anteriorly convex) parts of the sinuation; the part of the head behind the strongly punctured portion is finely and densely clothed with scales. The proportions of the prothorax scarcely differ from those of *L. Darwini*, except that the segment is slightly more transverse; the lateral margins are slightly crenulate, and behind the post-medial curve (which is almost angular) are very nearly straight, and the hinder angles quite obtuse; the surface is rather closely (especially towards the sides), but not coarsely, punctured and scaled; there is a smooth dorsal line in the hinder half; the front is very finely margined, the margin punctureless. The elytra and scutellum are punctured and scaled even more finely and closely than the prothorax, the former being wrinkled transversely, but not coarsely, and scarcely costate. The propygidium, pygidium, undersurface and legs scarcely differ from those of *L. Darwini*, except in the upper external tooth on the front tibiæ being stronger, and the apical tooth more strongly curved outwards. The apical ventral segment is more than half as long as the penultimate; a shallow channel runs down its middle and its hind margin is slightly emarginate in the middle.

Northern Territory of S. Australia; collected by Prof. Tate.

L. Rothei, Blackb.

This species was described by me in the "Transactions of the Royal Society of South Australia," of last year, but I was unable to state from what locality it had emanated. There have lately been referred to me for determination about half-a-dozen specimens of it taken in the Northern Territory, and no doubt the original type came from the same quarter. Comparing it with the two species described above, I find that the clypeus is somewhat intermediate, being less rounded at the sides and more transverse than in *L. delicatula*, but more rounded laterally and less transverse than in *L. Darwini*; the prothorax a little more transverse than in either and less rounded anteriorly, at its widest very little behind the middle; the elytra much more evidently costate, the puncturation throughout closer and the scales less conspicuous, &c., &c. I cannot find any sexual distinction among the specimens I have seen of *L. Rothei*, and believe them all to be females; and was probably mistaken in my conjecture that the original is a male. In all, the apical ventral segment is a little more than half as long as the penultimate, its surface glabrous and almost impunctate, and its hind margin gently and evenly arched.

[The length varies from $7\frac{1}{2}$ to 9 lines.

L. degener, sp.nov.

Fusco-ferruginea; subrugose confertim punctulata; supra sparsim squamoso-hirsuta; subtus densius distinctius squamosa; capite minus lato minus convexo; clypeo sat brevi, antice in medio fortiter reflexo-emarginato, ad latera rotundato; prothorace sat convexo; sat transverso, antice minus angustato, lateribus mox post medium ampliato-rotundatis, postice haud sinuatis, angulis posticis rectis; tibiis anticis acute tridentatis.　　Long. $6\frac{1}{2}$ lines.

The puncturation is on the whole rather uniform over the entire upper surface, but becomes gradually a little finer and less close from the clypeus hindward to the apex of the elytra ; the scales are small, elongate, and hair-like, and are rather closely distributed, but not very conspicuous. The clypeus is very strongly emarginate in front, the length from the base of the emargination to the clypeal suture being scarcely half the greatest length of the clypeus. The prothorax is about half again as wide as it is long down the middle, its base being less than half again as wide as its front margin. The elytra are scarcely costate, the indications of costæ appearing as mere irregularities of the surface which do not disturb the sculpture (in *L. Darwini* and *Rothei* the costæ are devoid, or very nearly so, of puncturation). On the underside the sculpture is of the same character as on the upper, but the puncturation is less rugose and more sparing, and the scales less hair-like. The teeth on the external edge of the front tibiæ resemble the same in *L. Rothei*, all being very sharp, the upper one much smaller than the lower two. The two specimens before me present no notable character likely to be sexual unless it be that the elytra of one are somewhat dilated behind the middle, while those of the other are narrower and more parallel.

The small size of this species will distinguish it from the three species previously described ; the very small hair-like scales of its upper surface and almost non-costate elytra also are distinctive.

Northern Territory of S. Australia ; taken by Mr. J. P. Tepper.

L. RUFA, sp.nov.

Rufa ; supra capillis brevibus sparsim, subtus squamis parvis confertim, instructa ; clypeo minus brevi, rugose crasse punctulato, antice in medio leviter reflexo-emarginato, ad latera rotundato ; prothorace fortiter subrugose nec crebre punctulato, sat convexo, sat transverso ; antice angustato, lateribus post medium ampliato-rotundatis, postice haud sinuatis, angulis posticis rectis ; elytris sat crebre minus fortiter punctulatis ; tibiis anticis obtuse tridentatis. Long. 6½ lines.

This species is closely allied to *L. degener*, from which it differs as follows : the clypeus is very much less strongly emarginate in front (resembling that of *L. delicatula*, but with the sides much less rounded), the prothorax is more narrowed anteriorly and much more sparingly punctured on the disc, the elytra are much more sparingly and less confusedly punctured, the scales on the upper surface are altogether hair-like, and the teeth on the outer margin of the front tibiæ are considerably blunter.

The three specimens before me appear to be of the same sex. The club of the antennæ is evidently longer than in the specimens of *L. degener* mentioned above (being nearly as long as the preceding six joints together), the elytra are dilated behind the middle, and the apical ventral segment is more than half as long as the penultimate (as in all the specimens of the genus that I have seen, except the one of *L. Darwini* mentioned above).

Northern Territory of S. Australia; taken by Mr. J. P. Tepper.

N.B.—It should be noted that in perfectly fresh examples the surface of some of the above species may very possibly be more densely scaled than I have described it, as the scales are very easily rubbed off; and it is probable that few cabinet specimens are quite as scaly as they were when freshly matured from the pupa.

Unless all the examples (with one exception) examined by me of the preceding species be of the same sex, the sexual distinctions are extremely slight.

PALMERSTONIA, gen.nov. (PIMELOPIDÆ).

Mentum in medio laminam compressam erectam conformans.

Mandibula prominentia, librata, extus obtuse bidentata.

Maxillæ haud observatæ.

Palpi maxillares validi, articulo 1° parvo, 2° subcylindrico, 3° parvo, 4° subconico antennarum clavâ paullo breviori.

Palpi labiales toti aperti, articulo 1° subcylindrico, 2° parvo, 3° subgloboso antennarum clavâ vix minore.

Antennæ 10-articulatæ, clava 3-articulata, sat parva.

Labrum breve transversum.

Clypeus ad perpendiculum directus.

Oculi magni.

Prothorax (? unius sexus solum) antice vix impressus.

Scutellum sat magnum transversum.

Tibiæ anticæ (? unius sexus solum) externe fortiter 3-dentatæ intus apice spinas singulas ferentes ; posteriores 4 unicarinatæ, apice spinas binas latas ferentes.

Tarsi antici perlongi, posteriores 4 breves, posticorum articulo primo valde dilatato.

Processus prosternalis post coxas erectus spiniformis.

Stridulationis organa nulla (?).

This is one of the most extraordinary insects I have ever seen. Its general appearance is that of a female *Pimelopus*, but some of its characters are quite anomalous, especially that of its exposed labial palpi, which are inserted on either side the vertical lamina of the mentum, and bear (as their apical joint) a large round ball flattened on one side. The clypeus is bidentate in front, very strongly angulated at the sides and thence narrowed to the base, which is roundly emarginate ; it is almost perfectly vertical, its plane being at right angles to the surface of the hinder part of the head, and its base rising considerably above the surface, so that viewed from the side the head seems to rise into a bilobed keel, and to be abruptly truncate in front,—while viewed from in front the clypeus looks like an erect shield. The mentum no doubt resembles that of *Nephrodopus*. The basal joint of the antennæ is not much shorter than the following six together ; the 2nd is considerably longer than any of the next five, which are all short ; the club is about as long as the 2nd joint of the anterior tarsi. The inner apical spine of the front tibiæ is a little longer than the basal joint of the tarsus. On the posterior 4 tibiæ there are traces of a transverse keel near the base in addition to the well-defined one below the middle. The front tarsi are very much longer than,

the middle equal to, the posterior shorter than, their respective
tibiæ. The hind legs scarcely differ from those of *Pimelopus*,
except in not having two distinct carinæ on the tibiæ and having
slightly longer tarsi with the 2nd joint not dilated. The prosternum in front of the anterior coxæ consists of an elongate
shining triangular plate, the wide end of which is directed forward,
the narrow end entering between the coxæ. The propygidium
is squamosely punctulate at the sides, delicately transversely
wrinkled in the middle; this transverse wrinkling may perhaps
be feeble means of stridulation. The pygidium bears a system of
very faint but not very fine puncturation which is more obsolete
in the middle than at the sides.

P. Bovilli, sp.nov.

Nitidus; valde convexus, latitudine majori pone medium posita;
totus rufo-castaneus; supra (clypeo labroque exceptis) glaber;
clypeo sparsim, prothorace vix evidenter, elytris sparsim obsolete
crasse, punctulatis; his oblique obsolete 4-costatis, striâ suturali
fortiter impressâ; subtus longe nec dense pubescens.

[Long. 12 lines, lat. 7 lines.

The entire insect is of a chestnut red color, a little paler on the
underside. The labrum is clothed rather thickly with long hairs.
The clypeus is nearly twice as wide as long, and is sparingly beset
with small sharply defined punctures, each of which bears a long
hair. The head behind the clypeus is very short. The prothorax
is about a third as wide again as long, its anterior margin gently
concave and about half the width of the base, the sides viewed from
above seeming to diverge very strongly from the small acute anterior
angles, and then run round with a very feeble curve to the base,
which is rather strongly bisinuate, the hind angles being scarcely
defined, but when viewed from the side the lateral margins are
seen in reality to form an even and strong curve; there is a
scarcely traceable impression at the middle of the disc near the front
margin, and in my specimen a series (very probably abnormal) of

four round smooth foveæ placed transversely across the disc a little in front of the middle. The scutellum is impunctate. The elytral costæ are all very faint, the external one scarcely discernible (in my example it is quite lost on one elytron, nearly so on the other). On the underside the following parts are clothed with long but not close pubescence,—the mentum, the front of the prosternum, the reflexed margins of the pronotum, the prosternal process, the lateral portions of the meso- and meta-sterna, and a line across each ventral segment. All the above pubescence is erect except that at the front of the prosternum, which is directed forward, but with a tuft in the middle erect. The legs, the basal two joints of the antennæ, and also the ocular canthus are clothed with long hairs.

A single specimen (apparently a female) was sent to me from Palmerston, N. Territory, by Dr. Bovill.

CACOCHROA OBSCURA, sp.nov.

Minus convexa ; nigra ; supra glabra ; pygidio, corpore subtus, et pedibus plus minus ferrugineo-hirsutis ; clypeo sat fortiter punctulato, antice sat fortiter emarginato ; prothorace sparsim obscure (lateribus sat fortiter) punctulato, angulis posticis rotundatis ; elytrorum disco fortius sublineatim punctulato, lateribus crasse transversim rugatis ; tibiis anticis 3-dentatis.

[Long. 8 lines, lat. 4 lines.

The head closely resembles that of *Cacochroa gymnopleura*, but is more closely punctured in the hinder part. The prothorax is gently lobed in the middle behind, and is not much less than twice as wide at the base as it is long down the middle, its front margin being less than half as wide as its base ; its hind angles are quite rounded off ; its surface is very sparingly and finely punctulate, except near the front margin (where the puncturation is rather stronger and closer) and near the sides (where it is very strong and coarse) ; the lateral furrow (within the thickened margin) does not reach the base, but ceases where the side begins to round off to the hind margin. The scutellum is elongate, with the apex sub-bifid owing to the presence of a longitudinal channel which commences obsoletely close to the base, and gradually deepens to the apex. The disc of the elytra is punctured uniformly with that of the prothorax, and also bears some much coarser puncturation, which has a tendency to run in rows, these rows seeming here and there to be placed in feeble striæ; the lateral portions of the elytra and also the apex are devoid of puncturation, but are sculptured with a well-developed system of coarse transverse wrinkles, commencing behind the post-humeral contraction; the apices are separately rounded; the suture is convex near the apex, but not at all produced behind. The pygidium is sparingly strigose and sparingly furnished with rather short hairs. The sternal portion of the undersurface is strongly and sparingly punctured (the flanks of the pro- and meso-sterna being strigose), the metasternum most strongly; of the ventral segments the first is transversely strigose on either side at the base, the rest are almost devoid of sculpture except that segments 2-5 are longitudinally concave in the middle (probably in one sex only) the concavity being punctulate and hirsute, that segments 4 and 5 have a transverse ciliated line of punctures on either side, and that segments 1-4 bear on either side a closely and finely punctured opaque space (very likely tomentose in a fresh specimen), which is subquadrate on segment 1 and triangular on the rest; there is similar sculpture on either side of the pygidium; the sterna are sparingly and shortly hirsute. The front coxæ and femora and the four posterior femora and tibiæ have their undersurface densely clothed with long pale hairs (? in both sexes), and the front tibiæ (perhaps in the female only) are tridentate, the upper tooth very much smaller than the others. The antennal club in the specimen before me (probably a female) is a little shorter than the length, in front of the eye, of the clypeus. The mesosternal process protrudes forward beyond the front of the intermediate coxæ nearly as far as the length of the basal 3 joints of the front tarsi, and is thick and somewhat cylindric at its base; in shape it resembles the same part in *Polystigma punctata*, Don.. but is longer and stouter.

The post-humeral contraction of the elytra is quite as in *Cacochroa*, and allows quite as much (as in that genus) of the metasternum and hind coxæ to be seen from above. The meso-thoracic epimera also are conspicuously visible from above.

This species does not appear to me to fall exactly into any of the numerous genera of Australian *Cetoniæ*, but, as it has very much the appearance of a *Cacochroa*, I have preferred to refer it to that genus, and describe its structural characters fully, rather than create a new genus. It differs from *Cacochroa* chiefly in having the base of the prothorax a little inclined to be lobiform in the middle (not so much as in *Polystigma*), and the hinder angles of the same rounded off, while the former of those characters, together with its deeply emarginate clypeus, &c., &c., separate it from *Aphanesthes*. The extremely strong post-humeral contraction of the elytra will distinguish it from most of the other genera which possess a long mesosternal process.

Northern Territory of S. Australia ; taken by Mr. J. P. Tepper.

BUPRESTIDÆ.

NEOSPADES LATERALIS, sp.nov.

Sat convexus ; capite prothoraceque læte viridibus, vix aureo-micantibus ; elytris obscure cupreo-æneis, albo-maculatis, antice lateribus læte viridibus ; subtus æneus vel viridis, abdominis lateri-bus albo-maculatis ; capite confertim fortiter rugose punctulato ; prothorace transversim strigoso ; elytris fortius vix crebre punctu-latis, apicem versus subtiliter serratis ; corpore subtus sparsim griseo-pubescenti, subtilius squamose (prosterno crasse fortius) punctulato ; tibiis posticis apicem versus tribus capillorum penicillis instructis.   Long. 4½ lines, lat. 1¾ lines.

The elytra are of a dull coppery æneous color. The spots of white pubescence are on each elytron—two on the hinder part of the lateral margin, and three or four near the suture in the hinder two-thirds ; the green patch commences on the base and extends slightly more than half way to the apex, occupying the external half of the surface but being interrupted by the humeral callus (which is of a golden copper color) and a little contracted at its apex externally ; it is bordered except at the base and the front part of its lateral edge by golden copper color. The pattern and colors of the upper surface do not appear to be variable. The underside is for the most part of a dull olivaceous tint, and is more or less closely covered with small scale-like pubescence ; in some examples the underside (especially the metasternum) is in parts of a decided green. Tne head is somewhat convex and is longitudinally impressed between the eyes. The prothorax is less than half again as wide as it is long down the middle, its base, which is bisinuate, being about a third as wide again as its front margin, which also is bisinuate ; the true margin is very slightly arched and is strongly bent under the head and front. The three little elevations (each bearing a pencil of pubescence) near the apex of the external margin of the hind tibiæ are a conspicucus character.

Very different from *N. chrysopygius*, Germ., in the markings of its elytra and other characters. *Cisseis cuprifera*, Gestro, is probably a member of this genus, and must be very close to Germar's species.

Northern Territory of South Australia ; collected by Mr. J. P. Tepper.

N. SIMPLEX, sp.nov.

Sat convexus ; nitidus ; viridis, aureo-micans ; elytris opacis, obscure cupreo-æneis, his et abdominis lateribus albo-maculatis ; capite sat confertim subrugose punctulato ; prothorace subtiliter transversim rugato evidenter punctulato ; elytris transversim strigosis, obscure punctulatis, apicem versus obsolete serratis ; corpore subtus sparsissime breviter griseo-pubescenti, subtilius squamose (prosterno crasse fortius) punctulato ; tibiis posticis postice nigro-ciliatis.   Long. 3½ lines, lat. 1½ lines.

The spots of white pubescence on the elytra and hind body are arranged as in the preceding species. The color of the elytra is very confused appearing different in different lights; it is a dull coppery purple, much brighter at the apex than in front, but in some lights the middle of the disc near the front and an ill-defined fascia just behind the middle appear blue; in some specimens the ground color appears to be brighter (especially towards the sides) than in others. The head is very decidedly more convex than that of *N. lateralis*, with the longitudinal furrow deeper and the puncturation not quite so close or rugose The prothorax scarcely differs from that of *N. lateralis* (apart from its color, which is a much more *golden* green), except in having somewhat more distinct puncturation in addition to its strigosity. The transverse wrinkling of the elytra is scarcely existent behind the middle; the puncturation is very obscure being rather sparing, coarse and shallow, the punctures in some specimens appearing to be filled up, and in some specimens the filling seems to protrude as though the elytra were obscurely granulose rather than punctured. The hind tibiae are lightly keeled externally in their apical two-thirds, the keel bearing a fringe of close-set erect fine black hairs.

Northern Territory of South Australia; collected by Mr. J. P. Tepper.

CISSEIS ELONGATULA, sp. nov.

Angusta; supra obscure cupreo-nigra, capite roseo-cupreo; subtus aenea sat nitida, elytris et prothoracis lateribus albo-maculatis; capite subplano, subopaco, longitudinaliter leviter sulcato, sat fortiter nec crebre nec rugose punctulato; prothorace transversim subtiliter strigoso; elytris squamose sat crasse punctulatis, antice et latera versus transversim rugatis, apicem versus serratis; corpore subtus obscure (prosterno crassius) squamose punctulato. [Long. 2½ lines, lat. ⅔ line.

The head is nearly flat with a distinct dorsal furrow, and is opaque of a bright rosy color, and its puncturation is rather large and moderately close, very clearly defined, but not deep or rugose, the punctures very little confused by transverse wrinkles; the head is very like that of *C. roseo-cuprea*, Hope, but more strongly punctured. The prothorax is a little more than a third as wide again as long ; the base and front margin are both bisinuate, the former about a quarter as wide again as the latter; the sides nearly straight, the surface very delicately wrinkled transversely with a fairly well-marked longitudinal impression on either side at the base. The puncturation of the elytra is very vague, and scale-like in appearance, the transverse wrinkles are fairly well-defined in front and at the sides. The spots of white pubescence on the elytra resemble those of the preceding two species. The almost flat, opaque, finely coriaceous head with distinct, not very close-set punctures, and scarcely any trace of wrinkles, together with very small size, narrow parallel form, and obscure color will distinguish this from all other species yet described of *Cisseis*. In my opinion it, *roseo-cuprea* and some other species might well form a new genus differing from *Cisseis* in their short strongly compressed tarsi, which approach those of *Neospades*, though the claws and antennæ resemble *Cisseis*. These insects seem to occupy a doubtful position between *Agrilidæ* and *Trachydæ*.

Northern Territory of S. Australia ; taken by Mr. J. P. Tepper.

TENEBRIONIDÆ.

HELÆUS.

Through the courtesy of the Hon. W. Macleay in examining a series of *Helæus* from my collection and comparing them with his types, as well as in furnishing me with types of several species that were not represented in my collection, I am able to offer some notes on this genus, together with descriptions of several new species, and to do so with some confidence that I mean by the various specific names the same insects that are referred to under those names in No. V. of the "Miscellanea Entomologica." I do not consider it a certainty that in every case Mr. Macleay applies the names to the same insects that were before the original describers,—nor does Mr. Macleay himself consider it so ; but as there are so many of these of which the positive identification is (either absolutely, or) at least to Australian students, *impossible*, I think Mr. Macleay has acted on the right principle (in dealing with those descriptions which might apply to anyone of several species) in selecting a particular one to bear the name and describing it so fully,—expanding the original description,—that it may be at least clear to what insect *he* applies the name,—thus leaving to any student who may possess information that has not come before Mr. Macleay the burden of correcting him if he is wrong. There are so many Australian species the types of which Europeans have described badly and then lost, that Australian students must choose between the course Mr. Macleay has adopted and that of holding aloof from describing the insects of their own country.

In dealing with the species before me of this genus I propose, then, to accept the whole of Mr. Macleay's determinations as conventionally correct, although in some instances I may express a doubt of their absolute correctness, so that my remarks will be in harmony with his valuable monograph of the genus, and consequently whatever corrections may eventually be applied to his determinations will have to be read into my remarks.

H. PRINCEPS.

I expect to find eventually, as Mr. Macleay evidently thinks probable, that the South Australian species to which he applies this name is distinct from the Western Australian species on which Hope's vague description is founded.

H. INTERMEDIUS, de Brême.

I regard this as the most doubtful of all Mr. Macleay's determinations, as I fail to find in the insect to which he applies this name any more distinct abbreviated elytral costa than there is in all its allies. In my opinion the following species (which Mr. Macleay considers hitherto undescribed) is quite as likely to be the true *H. intermedius* as that to which Mr. Macleay has appropriated the name, but, in accordance with the principle I have laid down, I accept his decision.

H. BREVICOSTATUS, sp.nov.

Sat latus; ovatus; minus nitidus; piceus, marginibus dilutiori-bus; prothoracis marginibus subtilissime nec crebre granuloso, disco cornu valido erecto instructo; elytrorum disco confuse crebre sat fortiter punctulato, sutura fortiter costata et utrinque costa abbreviata minus fortiter elevata instructa; marginibus subti-lissime sat crebre granulatis. Long. 14 lines, lat. 8⅓ lines.

This species resembles that which Mr. Macleay in his monograph describes as probably identical with *H. princeps*, Hope; compared with it the present insect is less dilated about and behind the middle of the elytra (the hinder part of the margin being very much narrower); the anterior prolongations of the prothorax are pointed and very much narrower, the turned-up edges of the same being thicker, less elevated and less erect; the disc of the elytra is devoid of granules (except the marginal row) and the margins of the elytra are furnished only with excessively fine granules scarcely larger than those on the margins of the prothorax. *H. brevicostatus* is distinguished from all its allies by the abbreviated costa commencing at the base of each elytron just outside the scutellum and running obliquely towards the suture; this costa is about 1½ lines long and, though decidedly less elevated than the suture, is perfectly well-defined.

The prothorax is twice as wide as its length (from the base to the apex of the anterior prolongations), the disc occupying more than a third of its total width, and being uneven and finely punctured; the "intermediate expansion" (as I will call the space between the disc and the turned-up edge that forms the true margin) is rather closely, evenly, and very finely granulate; the true external margin is very thick, and narrow but not vertical; the prothoracic horn is rather short, very stout, and scarcely directed backward at the apex. The intermediate expansion of the elytra is as wide at its base as that of the prothorax, but contracts rapidly to little more than half that width, and thence continues of somewhat even width to the apex;

it is much turned-up, and its sculpture does not differ much from that of the same part of the prothorax ; the true margin of the elytra is much wider than that of the prothorax, and is most erect at about a third of its length from the base.

A single specimen ; the locality of its capture is uncertain.

H. HORRIDUS, sp.nov.

Oblongo-ovalis ; convexus; nitidus ; ater ; prothorace sub-tilissime punctulato minute granulato ; elytris sat fortiter lineatim punctulatis, fortiter 7-seriatim tuberculatis.

[Long. 7 lines, lat. 4 lines.

The prothorax is at its widest at the base, whence its sides are contracted in a very gentle curve to the front ; the anterior projections are quite slender and slightly crossed ; the distance from the base to the front margin is decidedly less than half again as great as that from the front margin to the apex of the anterior projections ; the intermediate expansions are moderately wide, and are granulated not very finely, with some appearance of transverse wrinkling ; the margins are extremely thick and erect ; the disc is very finely and evenly punctured and also finely granulated, but there is a space almost free from granules on either side of the central line which is rather strongly keeled, the keel being strongest at the base ; the width of the prothorax is scarcely half again as great as its length. The elytra are densely punctured in 13 rows, which, however, are rendered scarcely traceable by the extent to which they are interrupted by the rows of large round tubercles that occupy the alternate interstices ; the 1st of these rows commences outside the scutellum as a strong interrupted costa which runs obliquely to the suture, and then margins it as a series of large tubercles nearly to the apex ; the 3rd row of tubercles also commences as an interrupted costa ; the tubercles of the 3rd and 5th rows are rather larger than the others and attain the apex ; in the 5th row there are about 10 tubercles ; the intermediate expansion is moderately wide at the base, but soon becomes narrower, and continues so to the apex, its surface being quite smooth except close to the base, where there are a few tubercles ; the margin is thick and well-defined : the epipleuræ of the elytra are very coarsely and strongly punctured. The underside is minutely granulate. The hind tibiæ are rather strongly flexuous (perhaps a sexual character) ; the prosternum is not at all carinate.

An extremely distinct species belonging to the same section of the genus as *H. echinatus*, Hope. Compared with that insect the following (among others) differences may be noted : general form very much narrower and more elongate ; anterior processes of the prothorax much more projected forward making the segment longer; intermediate expansion of both prothorax and elytra decidedly narrower, but at the same time more sharply defined ; the tubercles in the rows on the elytra in general much larger, especially those in the 1st, 2nd, 4th, 6th, and 7th rows, which, however, are very much smaller than those in the 3rd and 5th.

The South Australian Museum possesses a single specimen, probably taken in South Australia.

SARAGUS INÆQUALIS, sp.nov.

Ovalis ; minus opacus ; ferrugineus, capite prothoracisque disco infuscatis ; hoc minute granulato, marginibus reflexis ; elytris valde rugosis, antice tricostatis, interstitiis in parte postica tuberculatis, tibiis anticis calcare apicali gracili acuminato.

[Long. 7 lines, lat. 4¾ lines.

This species is so closely allied to *S. lævicollis*, Fab., that it will be sufficient to add to the above diagnosis an enumeration of its differences from that insect.   Its ferruginous color, with only the head and the disc of the pronotum and prosternum darker may not be constant.   Its shape is quite distinctive, the elytra being considerably longer in proportion to their width than those of *lævicollis*, and being uniformly, though very gradually, narrowed from the base to their apical half, which is rapidly contracted, the apical part being more pointed than in *lævicollis*.   The front part

56

of the intermediate expansion of the prothorax is distinctly con-cave owing to its being turned up at the edge. The general surface of the elytra is much more coarsely rugose than in *lævi-collis*, the costæ and tubercles being very similar when closely examined, but appearing at the first glance less conspicuous owing to the greater rugosities among which they are placed. The anterior tibiæ are narrower, the apical spur being long, very much more slender, and acutely pointed.

I have a single specimen taken by Mr. J. J. East near Mallala.

S. LINDI, sp.nov.

Late ovatus; opacus; niger; capite prothoraceque confertim subtiliter granulatis; hoc quam longiori multo plus duplo latiori, sat late marginato, marginibus planatis; elytris minute sparsius granulatis, fortiter tricostatis, interstitiis seriatim tuberculatis; tibiis anticis calcare robusto breviori instructis.

[Long. 6 lines, lat. 4½ lines.

Another ally of *S. lævicollis*, F., but very distinct from it. The prothorax at its widest is twice and a-half as wide as down the middle it is long, and its base is nearly twice and a-half as wide as its front, the margins being without transverse wrinkles, and the basal portion being declivous backward, and bearing a large well-marked central impression; in other respects the head and pro-thorax resemble those of *S. lævicollis*. The elytra are not longer than together wide; their general surface, underlying the granules tubercles and costæ is quite smooth; this surface is sprinkled with minute granules which are sparing about the region of the scu-tellum but become closer towards the margin and apex; the suture is scarcely elevated near the base but becomes strongly so on the hinder declivity, and is bordered on either side by a row of close-set large granules, or small tubercles, some of which are conical and some elongate; each elytron bears 6 rows of strong elevations which cease at the beginning of the hind declivity; the 2nd and 4th of these form undulated nearly uninterrupted costæ; the others consist of elongate ridges resembling disconnected portions of costæ; as they are not symmetrical on the two elytra of my specimen the degree in which they are interrupted is no doubt very variable; the elytra have rather strongly rounded sides and are widest in the middle; the intermediate expansion is considerably wider than in *S. lævicollis*, and is but little narrowed near the apex. The underside resembles that of *S. lævicollis*. The legs and antennæ are a little reddish. The apical spur of the anterior tibiæ is short and blunt.

A single specimen occurred to me at Port Lincoln.

S. LATUS, sp.nov.

Sat nitidus; subhemisphæricus; piceo-brunneus; capite prothoraceque confertim subtiliter granulatis; hoc quam longiori multo plus duplo latiori, sat late marginato, margine antico concavo; elytris minute granulatis, tricostatis (costis plus minus interruptis), interstitiis internis obscure seriatim tuberculatis; tibiis anticis calcare gracili acuminato instructis.

[Long. 6 lines, lat. $4\frac{1}{2}$ lines (vix).

The head and prothorax are closely and minutely granulated, many of the granules on the latter, especially about the middle, being much elongated so as to give an appearance of longitudinal wrinkling. The prothorax is quite twice and a half as wide as it is long down the middle, and its base is twice and a half as wide as its front margin; the intermediate expansion has no transverse folds, its anterior part is distinctly concave, and its granulation is quite continuous with that of the disc. The elytra are not at all narrowed at the base (which gives the insect a very distinctive subhemisphæric appearance); they are not at all longer than together wide; the lateral half of their anterior margin is very obliquely cut away so as to meet the lateral margin in a very obtuse angle; the intermediate expansion is very wide in front (considerably more so than in *S. lævicollis*), and is narrowed uniformly to the apex, where it is not wider than the same at the apex of the elytra of *S. lævicollis*, its surface being marked with transverse folds, and also with tubercles similar to those of the disc (in some specimens the tuberculation is very obsolete) ; the disc is furnished tolerably evenly with rather close-set granules or small tubercles : the suture is scarcely elevated in any part ; the three ridges usual in this section of *Saragus* are represented, the 1st by a strong straight (not undulating as it is in *S. lævicollis*) costa not reaching the apex, and having its apical half (or more) broken into tubercles,—the 2nd by a similar costa, which, however, is in most examples broken into tubercules from just behind its base,—the 3rd by a row of tubercles : each of the intervals between the suture and the 1st costa, between the 1st and 2nd and between the 2nd and 3rd costæ is occupied by a row of tubercles somewhat larger than those that form the general granulation of the surface, but there is no serial tuberculation whatever outside the 3rd costa. The spur of the anterior tibiæ is very much more slender and pointed than that of *S. lævicollis*. This species is not very close to any other *Saragus* known to me. Its subhemisphæric form and shining surface will distinguish it from most of the species with the anterior tibiæ strongly spurred.

Murray Bridge ; taken by Mr. Tepper.

S. MEDIOCRIS, sp.nov.

Subopacus : late ovatus ; brunneo-niger : capite prothoraceque confertim subtiliter granulatis ; hoc quam longiori plus duplo latiori, minus late marginato, margine antico vix concavo ; elytris minute granulatis, tricostatis (costis undulatis plus minus interruptis), interstitiis seriatim tuberculatis ; tibiis anticis calcare breviori obtuso minus gracili instructis.

[Long. $5\frac{1}{2}$ lines, lat. $3\frac{3}{4}$ lines.

The head and prothorax scarcely differ from the same parts in *S. Lindi*, except in the intermediate expansion of the latter being a little narrower. The elytra also resemble those of the same species almost exactly in respect of their costæ and rows of tubercles : there is, however, no well-defined row of tubercles running down the sides of the suture : the surface of the disc is very much rougher (making the small granules much less conspicuous) than in *S. Lindi* ; the elytra are very little narrowed at the base; and their intermediate expansion is extremely narrow even at the base, being markedly narrower than in *S. lævicollis*, with margins scarcely marked at all.

A single specimen in my collection; I have no note of the exact locality beyond that it was taken in South Australia.

SARAGUS MACLEAYI, sp.nov.

Late ovalis; convexus; sat nitidus; piceo-niger, antennis palpis et (nonnullis exemplis) tarsis dilutioribus; capite prothoracoque duplo-punctulatis (subtiliter et minus subtiliter); elytris multo fortius lineatim, interstitiis sparsim subtilissime, punctulatis, his nonnullis obscure convexis.

[Long. 4-5 lines, lat. 2⅔-3 lines.

The prothorax is considerably more than twice as wide as down the middle it is long, and its base (which is bisinuate) is more than twice as wide as its front margin (which is deeply emarginate); its front angles are obtuse, the hinder ones sharp; its intermediate expansion is on either side rather less than a quarter the width of the disc, is not horizontal but declivous (though not sufficiently so to continue the lateral declivity of the prothorax evenly), and is gently narrowed from the base to the apex; its true margin forms a well-defined shining edge; the disc (as also the head) is thinly furnished with extremely minute punctures and also with larger (but still fine) ones; there is a transverse impression close to the front which makes the middle of the anterior margin appear somewhat elevated; there are also a number of obscure impressions all across the base (the intermediate expansion is much more roughly sculptured); each elytron is furnished with about 17 rather irregular rows of moderately coarse punctures, the interstices between the rows (especially the 4th, 8th, and 12th) being obscurely convex; the intermediate expansion is wide at the base, then contracts rapidly, and then continues rather narrow, but almost even in width, to the apex. The apical spine of the anterior tibiæ is not particularly large.

This species must be allied to *S. brunnipes,* Brême, and must differ from it *inter alia* in having only a small spine at the apex of the anterior tibiæ ; from *S. brunnipennis* it differs *inter alia* in having its thorax more strongly punctured, and the interstices of its elytra more or less convex. Mr. Macleay tells me that it is distinct from everything known to him.

Sleaford Bay, near Port Lincoln.

S. ASPERIPES, Pasc.

An insect which I have taken several times at Port Lincoln agrees with Mr. Macleay's type (the Hon. gentleman informs me) of this species, and corresponds very well with Mr. Pascoe's description, but it should be noted that it is exceptional for the intermediate expansion of the elytra not to be marked with transverse folds. *S. asperipes* has much the appearance of a *Phosphuga.*

S. SATELLES, sp.nov.

Late ovalis ; sat convexus ; minus nitidus ; piceo-niger, marginibus dilutioribus ; capite subrugulose sat fortiter, prothorace duplo (subtiliter et subtilissime), elytris vix seriatim minus fortiter nec crebre, punctulatis ; his obsolete tricostatis.

[Long. 7¾ lines, lat. 5 lines (vix).

The head is rugosely, confusedly, and closely punctulate with very fine and rather coarse punctures intermingled. The prothorax is twice and a half as wide as it is long down the middle, its base a little more than twice as wide as its front margin ; its intermediate expansion is on either side about a third the width of the disc, and is subhorizontal, nearly flat, and of somewhat even width throughout its length ; its true margin is well-defined, thick, shining, and not erect ; the lateral outline is well rounded, the greatest width of the segment being just in front of the base ; its surface is punctured on a uniform system, the puncturation is stronger and rougher close to the lateral edges (except near the posterior angles), and becoming gradually smoother and more sparing towards the middle of the disc, and consists of fine and very fine punctures intermingled, the coarsest part being decidedly less coarsely punctured than the head. The punctures on the elytra are feeble (not much larger than the largest of those on the head) ; on the front part of the disc they run in traceable rows, but are much confused towards the apex and margins ; there is no striation on the elytra ; there are three longitudinal spaces representing the 4th, 8th, and 12th interstices, which are faintly convex and quite devoid of puncturation; the intermediate expansion is separated from the disc by a row of much larger punctures, and is rather narrow at the base, but does not contract much hindward, being at the apex about half as wide as at the base ; at the middle of its length it is about as wide as the interval on the front part of the disc of the elytra between two of the rows of punctures.

This species belongs to the *Phosphuga*-like group of *Saragus* ; compared with *S. asperipes*, Pasc., it is much larger, with the elytra not at all striate and much more finely punctured, the intervals between the rows (where they are traceable) of punctures being very much wider, the humeral angles quite rounded off, and the hind tibiæ devoid of distinct hispid asperities. Mr. Macleay has done me the favour of comparing the species with the types of *Saragus* in his collection, and does not find it identical with any of them.

Port Lincoln.

SARAGODINUS TUBERCULATUS, sp.nov.

Ovalis; o ·acus ; ater ; antennis pedibusque plus minus picescentibus ; capite prothoraceque confertim subtiliter, elytris sparsim seriatim, tuberculatis ; tibiis anticis externe haud crenulatis.

[Long. 5¾-6½ lines, lat. 3½-3¾ lines.

The clypeus (which is not separated from the rest of the head in any defined manner) is concave, especially towards the sides, the margins not defined, the front widely and gently emarginate.

The entire head is covered with small tubercles which are confused and obscure on the clypeus.   The prothorax is nearly twice as wide at its widest part as down the middle it is long, and its base is about two-thirds again as wide as its front margin ; its margins are sinuately divergent from the front to slightly behind the middle, where they are strongly and abruptly rounded, and then with a strongly sinuated curve converge to the base, but in such manner that they are nearly parallel close to the base and the whole prothorax has a cordate appearance ; the front margin is very strongly emarginate, the anterior angles well defined ; the base is scarcely bisinuate, the hind angles small, acute and directed obliquely outward and hindward ; the disc is strongly convex, the lateral margins wide (together more than half the width of the disc) and very strongly reflexed ; the surface of the entire segment is confusedly covered with tubercles which are very small and obscure towards the sides, but on the disc are considerably larger and more shining and sparing ; the lateral edges are strongly crenulated.   The scutellum is situated at the bottom of a depression in the elytra.   These latter are not quite a quarter as long again as together wide, and are evenly and gently rounded laterally (the humeral angles quite rounded off) ; each of them bears four rows of strong, slightly shining, conical tubercles (about 7 or 8 tubercles in each of the inner two rows, about 5 or 6 in the next, and about 10 in the outer one, which is close to the margin) ; some of the tubercles are larger than others, but the large and small ones are pretty evenly distributed along each row ; the tubercles have a little tendency to an elongate ridge-like form close to the base, and those of the outmost row are mostly a little smaller than the rest ; the spaces between the rows of tubercles and between the tubercles in the rows is all uniformly rugose and finely but not closely punctured ; the suture of each elytron is thickened and crenulate ; between this and the first row of tubercles, and also in each interval between the rows of tubercles are a few very small tubercles ; there is no defined line separating the upper surface of the elytra from their epipleuræ (which are strongly punctured), but a fairly distinct thickening of the margin (most noticeable from beneath) divides them. The general style of puncturation on the underside consists of well-defined coarse punctures, each puncture containing a kind of granule on which is a golden seta; on the lateral parts of the prosternum the granules protrude, giving the appearance of tubercles; on the rest of the undersurface the punctures are rather feeble and sparing down the middle becoming deeper towards the sides; those about the sides of the sterna are the best developed, and show the golden setæ most conspicuously. The legs are punctured and clothed with short inconspicuous hairs; the tibiæ are straight, the anterior having a strong tooth externally, near the apex, in addition to the large robust apical spur. The prosternal process is horizontal and slightly prominent behind; it continues backward beyond the hind level of the coxæ and its hinder declivity is almost perpendicular.

This species must be somewhat closely allied to *S. Duboulayi*, Bates, but besides being much smaller differs *inter alia* in the elytral sculpture, which is devoid of any distinct costæ, the intervals between tubercle and tubercle in each row being quite continuous with those between the rows of tubercles.

I obtained two specimens of this insect dead, but very little damaged, in a spider's web under a log about twenty miles north of Port Lincoln.

www.ingramcontent.com/pod-product-compliance
Lightning Source LLC
Chambersburg PA
CBHW022003190326
41519CB00010B/1372